CHALLENGE

Suggested Activities to Motivate the Teaching of Mathematics in the Intermediate Grades

AUTHOR

Mary E. Platts

**PUBLISHED BY
EDUCATIONAL SERVICE, INC.
P.O. Box 219
Stevensville, Michigan 49127**

TABLE OF CONTENTS

CHAPTER I: "DEVELOPING BASIC SKILLS"   PAGE

CHAPTER II: "GAMES FOR ONE OR TWO"

iv

CHAPTER III: "PUZZLES AND BRAIN TEASERS"

CHAPTER IV: "ARE YOU REALLY THINKING?"

CHAPTER V: "GEOMETRIC ART"

CHAPTER VI: "MISCELLANEOUS"

INTRODUCTION

Mathematics is a skill that has allowed some people to create timeless works of art, to build bridges and skyscrapers and even to walk on the moon.

It is also a subject that has forced some people to gnash their teeth and tear their hair and vow they'll **never** learn the 9's table or how to figure the square root of 256.

Teaching methods have an extremely important influence on the attitude of children towards mathematics. Drill is necessary to develop accurate skills, but drill can be interspersed with activities that allow children to see the subject also has its creative and thoroughly enjoyable side. CHALLENGE is dedicated to exposing children to that creative and enjoyable side of mathematics.

Each activity includes a description of all necessary preparations and materials, and an example of exactly how to present that activity to your group. The material was condensed as much as possible in order to include a maximum number of activities within the space limitations of a truly functional handbook.

CHAPTER I:
"Developing Basic Skills"

Activities to develop understanding and accuracy in basic mathematical skills.

PLACE VALUE (Grades 4-6)

A. Preparation and Materials: Write on the board the example shown below. Children will need paper and pencils.

Example:

1. 4 3 6 2
   a. _____
   b. _____

2. 9 1 6
   a. _____
   b. _____

3. 1 8 3 4
   a. _____
   b. _____

4. 8 9 9
   a. _____
   b. _____

5. 1 3 6 2
   a. _____
   b. _____

6. 1 4 3 2
   a. _____
   b. _____

7. 6 2 5 8 4
   a. _____
   b. _____

8. 2 3 1
   a. _____
   b. _____

9. 7 2 1 3 9
   a. _____
   b. _____

10. 9 8 6 9 8 6
    a. _____
    b. _____

B. Introduction to the Class: Look at the first set of numbers. How could you arrange those four numbers to name the **highest** quantity possible? Yes, Tami, 6432. Why is this so? (Putting the highest numbers in the columns of highest place value gives the highest total value.) Write 6432 on line a.

How could you arrange these same four numbers to name the **lowest** quantity possible, June? Yes, 2346. Why is this so? (Putting the highest numbers in the columns of lowest place value gives the lowest total value.) Write 2346 on line b.

Complete the others in the same way. Put the combination of highest value on line a. Put the combination of lowest value on line b.

*This activity is available in Challenge Volume I of the **Spice**™ Duplicating Masters.

2. MATCH ME (Grades 4-6)

A. Preparation and Materials: Cut as many flashcards as you have children in the room. Use these to make pairs of cards, one giving a mathematics vocabulary term and the other showing an example of that term. Adapt the vocabulary to your grade level ability. In the examples below, the circles indicate the specific part to be considered.

Example:

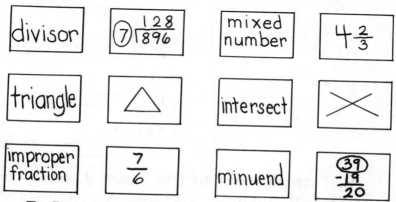

B. Introduction to the Class: We're going to play a mathematics matching game. Please count off by two's and form two lines facing each other. (If the class does not divide evenly, let the "odd" child be scorekeeper.)

To begin the game Team A will hold the vocabulary cards (show one to the class) and Team B will hold the examples (show one). Each person on each team will hold one card.

The first person on Team A will hold up his card and say to Team. B, "Match Me." The person on Team B whose card matches will hold it up. If he is correct, Team B scores one point. If he is incorrect or if anyone on Team B holds up a card that doesn't match even though the correct card is also being shown by a teammate, no point is scored.

Notice that during this part of the game only Team B has a chance to score. After each person on Team A has held up his card, we will collect the cards, shuffle them and the teams will trade cards. During this next part of the game Team A will have a chance to score.

When both teams are finished, the team with the highest score will be the winner.

*3. BRIDGING (Grades 4-6)

A. Preparation and Materials: Each child will need a sheet of ruled paper and a pencil. Place on the board the chart shown below, putting the numbers in row 1 across.

Example:

On Board

46	18	25	31	17	43	24	32
49							
52							
55							
58							
61							
64							
67							

Child's Paper

46	18	25	31	17	43	24	32
49	26	29	40	26			
52	34	33	49	35			
55	42	37	58	44			
58	50	41	67	53			
61	58	45	76	62			
64	66	49	85	71			
67	74	53	94	80			

B. Introduction to the Class: Today, I want to see how quickly and accurately you can bridge the decades in adding. First, fold your paper in eighths, making the folds go the opposite way as

—5—

the lines on your paper. At the top of each of the 8 columns write the number you see on the board.

Now I will give you directions for adding. In column 1 add 3 to 46. Then add 3 to your answer. Continue adding 3 to each answer until you have 8 numbers in all in column 1 down. Work quickly and comfortably, but DO NOT count fingers, objects or dots. Make yourself THINK THE ANSWERS.

(Allow children time to complete column 1. Then give directions for column 2. When that is completed by the children, give directions for column 3, and so on. Sample directions might be:

In column 2 repeatedly add 8 to each answer.

In column 3 add 4 to each answer.

In column 4 add 9 to each answer.

If this activity is used often, you might want to allow children only so many seconds to complete each column. Each time they do this exercise they could try to work faster while still maintaining accuracy.)

4. QUICK QUIZ (Grades 4-6)

A. Preparation and Materials: Ask children to draw a grid, like the one shown in the example, on paper. Above the top of the chart have them number the sections from 1 to 9. Down the side of the chart have them number the sections from 1 to 9 also.

If you use this activity often, it is practical and convenient to duplicate a generous supply of these charts.

B. Introduction to the Class: You are going to use these charts to show me how well you know your multiplication tables. Find the number 1 on the left side of your chart. In that

*This activity is available in Challenge Volume I of the Spice™ Duplicating Masters.
**This activity is available in Challenge Volume II of the Spice™ Duplicating Masters.

row across you will write the answers to 1 x 1, 1 x 2, 1 x 3, and so on (demonstrate).

Example:

Original Drawing   Completed Chart

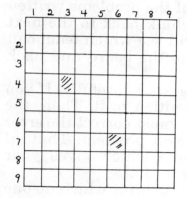

	1	2	3	4	5	6	7	8	9
1	1	2	3	4	5	6	7	8	9
2	2	4	6	8	10	12	14	16	18
3	3	6	9	12	15	18	21	24	27
4	4	8	12	16	20	24	28	32	36
5	5	10	15	20	25	30	35	40	45
6	6	12	18	24	30	36	42	48	54
7	7	14	21	28	35	42	49	56	63
8	8	16	24	32	40	48	56	64	72
9	9	18	27	36	45	54	63	72	81

Where would you write the answer to 3 x 4? Yes, David, in the square where row 3 across and column 4 down meet. (Note that square is shaded in the example.)

Where would you write the answer to 6 x 7? (The square is shaded in the example. Proceed in this way until you are sure the children understand how to use the charts.)

I will give you 5 minutes to complete your charts. (Adjust the time limit to the ability level of your group.) At the end of that time we will check your work together.

5. WHO'S THE WINNER? (Grades 4-6)

A. Preparation and Materials: Write on the board as many mathematics problems as there are children in your class. Divide the class into two equally-matched teams. Give the first person on each team a piece of chalk and an eraser.

B. Introduction to the Class: When I say, "Go," the first person on each team will go to the board, work any one of the problems written there and then take the chalk back to the next person his team. We will see which team can have all its members up to the board and back again first.

Watch the other team's work carefully. If you see an error call their attention to it. Then, the next member of their team must correct the error before he may work his own problem. That, of course, takes valuable time which gives your team a better chance of winning.

If the other team makes an error which no one spots before their next player begins working, they do not need to correct that error. So you can see it is to your team's advantage to be alert to your opponent's mistakes!

Are you ready? Go!

6. RHYTHM (Grades 4-6)

A. Preparation and Materials: None.

B. Introduction to the Class: Watch and listen carefully to the rhythm pattern I will show you. When you think you are ready, you may join me in this pattern.

(The pattern consists of clapping your hands twice, snapping the fingers of the right hand, then the left hand. These four beats are done in a steady rhythm over and over again:

Clap, clap, snap, snap.
Clap, clap, snap, snap.

When all the class has joined you in this rhythm pattern, proceed.)

Now, I will continue to clap and snap as before, but each time I snap my fingers I will give a mathematics problem. The first child in row one will clap as usual and as he snaps his fingers, he will give the answer to that problem and so on.

Let's see how far we can go around the classroom before someone breaks the rhythm.

Example:

TEACHER: Clap, clap, nine sevens.
CHILD 1:   Clap, clap, sixty-three.
TEACHER: Clap, clap, four eights.
CHILD 2:   Clap, clap, thirty-two.

C. Variation: Children may dictate facts to one another as they stand in rows or circles:

TEACHER: Clap, clap, five sevens.
CHILD 1:   Clap, clap, thirty-five.
              Clap, clap, six threes.
CHILD 2:   Clap, clap, eighteen.
              Clap, clap, four nines.
CHILD 3:   Clap, clap, thirty-six.
              Clap, clap, seven two's.

If this variation is used, the child who misses either by breaking the rhythm or by giving an incorrect answer must sit down. The child standing longest becomes the winner.

*7. MULTIPLE SEARCH (Grades 4-6)

A. Preparation and Materials: On the board write several rows of figures. In each row place in sequence one or more pairs of numbers which are multiples of a specific multiplication table you wish to stress. In the example shown,

*This activity is available in Challenge Volume I of the Spice™ Duplicating Masters.

the pairs which are multiples of 8 are circled for your convenience. Omit these circles when placing the figures on the board. Children will need pencils and paper.

Example: Multiples of 8

1. (2 (4) 8) 6 (3 2) 6 1 5 (7 2) 9

2. (9 (6) 4) 3 (1 6) 5 8 2 7 3 5

3. 4 3 (4 (8) 8) 2 7 (5 6) 2 (9 6)

4. 7 6 5 (8 0) 4 (4 0)(6 4) 2 1

5. 3 (1 6)(7 2) 2 5 1 7 (5 (6) 4)

6. (7 2) 3 6 5 (3 (2)(4)(8) 0) 1 9

7. 1 (5 6)(8 8)(4 0) 5 7 (3 (2) 4)

8. (5 (6)(4) 0) 3 9 7 8 (9 6) 2 5

9. (8 8) 4 4 (4 0) 1 (1 6)(3 2) 7

10. 4 (4 8)(9 (6) 4) 3 6 (7 (2) 4) 1

B. Introduction to the Class: In each row of numbers on the board there is at least one pair of numbers placed side by side that are an even multiple of 8. I want to see how quickly you can find these pairs.

Number your paper from 1 to 10. Beside number 1 write the pair or pairs of numbers in that row that are an even multiple of 8. Complete the other rows in the same way.

I will give you three minutes to complete your

work. (Adjust the time limit to the ability level of your group.)

You may begin now. (After the time limit is up, check papers together in class.)

8. TRAVELING (Grades 4-6)

A. Preparation and Materials: You will need a set of mathematics flashcards or a list of problems to be given orally.

B. Introduction to the Class: Everyone enjoys traveling. Today, you may travel all around the classroom **if** you know your mathematics facts.

The first child in row 1 will stand beside the second child in that row. I will hold up a flashcard. The one who correctly answers first may move on to the next desk while the loser sits down. You may keep moving from desk to desk as long as your answers are correct and you give your answer before your opponent does. In case of a tie, I will show a second card.

Mary, you may be our scorekeeper. As each child stands up, write his name. After his name write the number of desks he moved before someone took his place. Then, when our playing time is up, you can tell us who moved the greatest number of times. That person will be our champion traveler.

9. SOLITAIRE (Grades 4-6)

A. Preparation and Materials: You will need flashcards for reviewing mathematics facts at your grade level. Each card must have the correct answer on its reverse side.

B. Introduction to the Class: Here is a game you may play alone in your free time. Take this pack of flashcards to your desk. Place them in a stack, face up. Look at the first problem and think of its answer. Then, turn the card over to see if you are correct. (Insist that children **always** check, and never take it for granted that their answer is correct. This is to prevent them from teaching themselves a wrong answer.)

If your answer was correct, put the card to the right on your desk. If it was incorrect, put it to the left.

After you have gone through the entire stack, take the pile on the left of your desk and go through it in the same way. When you get all the cards moved to the right of your desk, shuffle the set and see if you can go through the pack and put them all on the right side of your desk on your first trial.

When you can answer all cards correctly, try timing yourself for speed. How quickly can you answer all the problems on the set of cards and still be accurate in your answers?

10. FLASHCARD DAY (Grades 4-6)

A. Preparation and Materials: Have each child in the room make sets of flashcards to review the basic addition, subtraction, multiplication and division facts, or to review multiplication or division of fractions, reducing fractions, finding common denominators, etc. These cards can be cut from tagboard or paper. Use a rubber band to hold each set of cards.

Make sure each card has the correct answer on its reverse side. Have each child make a chart on a 3 x 5 inch index card on which he may record a series of dates and scores.

Example:

Mary Jane	9/22	10/14	11/3	12/12	1/17	2/25	3/9	4/16	5/13
1. Addition	15	19	21						
2 Subtraction	12	13	15						
3. Multiplication	17	17	18						
4. Division	16	17	22						

B. Introduction to the Class: Each of you has your flashcards ready so it looks like we are all ready for Flashcard Day.

You may move your desks so that you can more easily work in pairs. When I say, "Go," one of you may hold your flashcards for your partner to answer one at a time. You can see the answer on the back of each card so you will know if he is correct. Put each card he answers correctly in one pile. Put the cards he answers incorrectly in another pile.

At the end of two minutes I will say, "Stop." (Raise or lower the time limit to fit the ability level of your group.) Count the cards you answered correctly and record this number and today's date on your chart.

Then, you may trade places so the other partner answers the problems.

(Flashcard Day becomes eagerly anticipated once it is begun as a periodic activity of the class. The facts are studied independently because of the competition. Also, each child gains encouragement by the improved scores which mark his progress.)

11. OVER THE RIVER (Grades 4-6)

A. Preparation and Materials: You will need a set of flashcards showing problems suitable for your grade level. Divide the class into two evenly matched teams which may stand on opposite sides of the room.

B. Introduction to the Class: Let's pretend that the two teams are on opposite sides of a river. I will show a flashcard. The first person on each team will compete against each other. The one who gives the correct answer first may stay on his own side of the river. The loser must cross over the river and go to the end of the opposite team's line. So, of course, each time your team answers last you are going to lose one of your team members.

Next, I will show a flashcard for the second person in each line and so on. The team with the most members when our playing time is over will be the winner.

12. FRACTION TYPES (Grades 4-6)

A. Preparation and Materials: Divide the blackboard into two sections. In each section list a group of fraction problems suiting the difficulty of the problems to the ability level of your group. Also, prepare a sheet containing examples of proper and improper fractions, mixed and whole numbers, mixing samples of types in no special sequence and giving many examples of each.

Divide the class into two teams.

B. Introduction to the Class: I will give a number to the first person on Team A. He must tell me if that number is an example of a proper or improper fraction, a mixed number or a whole number. If he answers correctly, he may go to the board and work one problem on his team's section of the board. If he does not answer correctly, he may not work a problem on the board.

Then, I will give a number to the first person on Team B and so on. At the end of the playing time we will correct the problems on the board. The team with the most correct answers will be the winner.

*13. WHICH ARE THE SAME? (Grades 4-8)

A. Preparation and Materials: Write on the board pairs of numbers, some of which match exactly and some of which differ slightly. In the example shown, the circles indicate pairs which match exactly. This is for your convenience in checking papers. Omit these circles when copying the work on the board. Children will need paper and pencils.

Example:

1.	658	685
(2.)	245	245
(3.)	32815	32815
(4.)	4486	4486
5.	201859	201589
(6.)	46894	46894
(7.)	678543	678543
(8.)	2198447	2198447
(9.)	853214	853214
10.	4580066392	458006392

B. Introduction to the Class: Have you ever missed a mathematics problem because you copied the problem incorrectly from the book? Here is an exercise that will give you practice in noticing likenesses and differences in number groups. This

*This activity is available in Challenge Volume I of the Spice™ Duplicating Masters.

proofreading skill will help you copy numbers exactly as you see them in the book.

On the board are pairs of figures. Some pairs are exactly alike and some pairs are slightly different. Look at each pair of numbers carefully. DO NOT COPY THE FIGURES. Just write the problem numbers of the pairs which are exactly alike. I will give you two minutes to complete your work. Ready? Go!

14. MENTAL MULTIPLICATION
(Grades 4-8)

A. Preparation and Materials: Draw on the board a chart like that shown in the example. Adapt the difficulty of the multiplication problems to suit the ability level of your group.

Example:

X	36	41	75	52
3				
5		X		
6				
2				
7				

B. Introduction to the Class: Please put away all pencils and paper. We're going to have an exercise in Mental Multiplication.

I will point to an empty square in this chart and call someone's name. That person will **mentally** figure the product of multiplying the number in that row down by the number in that row across.

For example, if I pointed here (point to the x in the chart) what numbers would you multiply together, Jenny? Yes, 41 x 5. You must do the multiplication **in your head.** Let's see what good mental mathematicians you can be.

C. Variation: Use the same kind of chart for mental calculations in division, subtraction or addition.

LATTICE MULTIPLICATION (Grades 4-8)

A. Preparation and Materials: None. Place examples on the board as directed.

B. Introduction to the Class: I would like to teach you a different way to multiply. Watch carefully, because after I have shown you this example. I will see how many of you can use this new method on other problems.

Suppose you wanted to multiply 53 by 27. Watch how I set up the problem. First I will put down this much:

Then I will add a diagonal line to each square:

Now I will multiply 3 by 2. The answer goes in the box under the 3. Since 6 is a one-digit number, I will use "0" in the ten's place:

*This activity is available in Challenge Volume I of the **Spice**™ Duplicating Masters.

Now I will multiply 5 x 2:

See where I put the answers to 3 x 7 and 5 x 7?

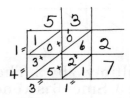

To find the answer I need only add each diagonal column. Begin adding at the lower right hand side and work around towards the left. Carry numbers into the next column when necessary. To read the answer, begin at the top left side and work down and across towards the right. Does this answer agree with the answer you get when you multiply the regular way?

Try Lattice Multiplication on these problems:

48	76	81	258	345	425	604
× 28	× 32	× 43	× 18	×24	×231	×460

A. Preparation and Materials: Use this activity only after children thoroughly understand the Lattice Multiplication procedure as described in the previous exercise.

Cut eleven strips of tagboard and number them as shown in the example.

Example:

0	1	2	3	4	5	6	7	8	9	Multiplier
0/0	0/1	0/2	0/3	0/4	0/5	0/6	0/7	0/8	0/9	1
0/0	0/2	0/4	0/6	0/8	1/0	1/2	1/4	1/6	1/8	2
0/0	0/3	0/6	0/9	1/2	1/5	1/8	2/1	2/4	2/7	3
0/0	0/4	0/8	1/2	1/6	2/0	2/4	2/8	3/2	3/6	4
0/0	0/5	1/0	1/5	2/0	2/5	3/0	3/5	4/0	4/5	5
0/0	0/6	1/2	1/8	2/4	3/0	3/6	4/2	4/8	5/4	6
0/0	0/7	1/4	2/1	2/8	3/5	4/2	4/9	5/6	6/3	7
0/0	0/8	1/6	2/4	3/2	4/0	4/8	5/6	6/4	7/2	8
0/0	0/9	1/8	2/7	3/6	4/5	5/4	6/3	7/2	8/1	9

B. Introduction to the Class: You enjoyed learning Lattice Multiplication, and here are some Lattice Multiplication Strips to make the job even easier.

Let's use the strips to multiply 625 times 3. Take the strips labeled 6, 2 and 5 and place them side by side. Place the multiplier strip on the right. Make sure all strips are in an even line with each other. Find the multiplier number 3 and look at the rows of numbers in the horizontal row to the left of the three.

Example:

*This activity is available in Challenge Volume I of the Spice™ Duplicating Masters.

Add these numbers diagonally just as you did in Lattice Multiplication.

Example:

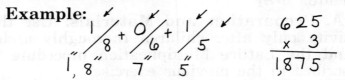

$$
\begin{array}{r}
6\,2\,5 \\
\times\ \ 3 \\
\hline
1{,}8\,7\,5
\end{array}
$$

(Let children try several problems using a single digit multiplier before proceeding.)

Now, let's use a two-digit multiplier. Let's try 935 x 27. Arrange the strips 9, 3, and 5 in a line with the multiplier strip just as you did before.

Example:

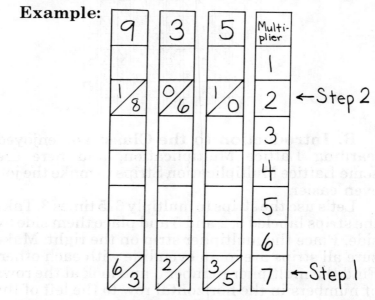

←Step 2

←Step 1

Step 1: Add the diagonals for the multiplier's one's place digit, which is 7.

Step 2: Add the diagonals for the multiplier's ten's place digit, 2, adding a zero to hold the one's place.

Example:

$$\overset{1}{\diagup} \underset{8''}{\overset{\diagup}{8}} + \overset{0}{\diagup} \underset{7''}{\overset{\diagup}{6}} + \overset{1}{\diagup} \underset{0''}{\overset{\diagup}{0}}$$

1,"8" 7 0" 0

Add zero in one's place.

Step 3: Add the two numbers to find the product:

		Proof
7 x 935 = 6545		935
+		x27
20 x 935 = 18700		6545
		18700
25,245		25,245

17. MULTIPLE MAGIC (Grades 4-8)

A. Preparation and Materials: None. Place examples on the board as directed.

B. Introduction to the Class: Here is a quick way to multiply any two-digit numbers between 10 and 20. For example, let's multiply 12 x 18.

1. Add the complete first number to the final digit of the second number. This tells you the number of tens in the answer.

Example: 12 + 8 = 20 tens = 200

2. Multiply the final digits of each number. This tells you the number of ones in the answer.

Example: 2 x 8 = 16 ones

3. Add the tens and ones together to discover the final answer.

Example: 200 + 16 = 216

Here are a few more examples:

1. 15 x 19 = ?
   15 + 9 = 24 tens = 240
   5 x 9 = 45 ones
   240 + 45 = 285

2. 13 x 18 = ?
   13 + 8 = 21 tens = 210
   3 x 8 = 24 ones
   210 + 24 = 234

Now, see if you can work the following problems using this new method. Check each one by regular multiplication to make sure your answer is correct. Which method is easier for you?

16 x 112 = ?   19 x 18 = ?
14 x 15 = ?   17 x 13 = ?

18. TURN A CARD (Grades 4-8)

A. Preparation and Materials: Cut forty-eight tagboard flashcards. On each card write a mathematics problem suitable for your grade level. The cards may show multiplication, addition, subtraction or division problems.
They may also show problems with fractions.

B. Introduction to the Class: Here is a game two to four of you may play in your free time. Shuffle the cards and divide them equally among all the players. Each player will put his cards face down in a pile in front of him.

Each will take his top card and lay it face up on the table. Each player will work the problem shown on his own card and announce the answer to the group. The player whose problem gives the highest answer wins all the cards face up on the table. If two or more cards give the same answer, no cards are won and new cards are played. When you win a set of cards, place them face down on the bottom of your pile.

Then, each player will turn up his next card and so on. At the end of the playing time the person with the most cards is the winner.

C. Variation: This game lends itself for drill in almost every area of mathematics. For example, you might make a set of cards in which each shows a Roman numeral. The highest numeral wins the face-up cards on the table. The cards could show problems in percentage, measurement or fractions. Have the children find the common denominator for the face-up cards and after converting all fractions to that denominator decide which fraction has the highest value. The person holding that card wins all face-up cards.

**19. MAGIC SQUARES (Grades 4-8)

A. Preparation and Materials: On scratch paper draw a square. Divide the square in fourths and each section in fourths again. Place numbers in each square so that each row across and down will add to the same total. In each row but one, circle one or two numbers. Keep this paper as your answer sheet.

Copy the magic square on the board. Put in all numbers except those which you circled on your answer sheet. Children will need pencils and paper.

—23—

Examples:

Grade 4 (Totals of 30)	Grade 5 (Totals of 622)	Grade 6 (Totals of $29\frac{1}{4}$)

6	(14)	7	3
(3)	1	6	20
7	(12)	6	5
14	3	11	(2)

360	52	(181)	29
138	(93)	240	151
(42)	325	(86)	169
82	(152)	115	273

$(5\frac{1}{2})$	6	$9\frac{3}{4}$	(8)
7	$3\frac{3}{4}$	8	$10\frac{1}{2}$
$9\frac{3}{4}$	(4)	$(9\frac{1}{2})$	6
7	$(15\frac{1}{2})$	2	$(4\frac{3}{4})$

B. Introduction to the Class (Using the first example): You are going to work with Magic Squares. A magic square adds to the same total in each direction; that is, the numbers in each row across and each row down will give you the same sum.

Notice there are some numbers missing. Our problem is to decide what number can go in each empty space to complete this magic square.

First, you must determine what the total will be. Do you see any row across or down that has all four numbers given? Yes, row 3 down. What is the total of that column? Good, Sarah, 30. Now can you make all the columns add to 30?

To find the missing number, total figures given and subtract from 30. What number goes into the space on column 1 down, Paul? Yes, 3.

Now look at column 2 down. What is different about this? Yes, Carol, there are **two** numbers missing. But who can see an easy way to find these two missing numbers? Yes, Douglas, you can work columns 1 and 3 across to find these numbers.

Let's see how many of you can find all the remaining missing numbers to complete this magic square.

(If you wish to make new squares for further drill, simply add or subtract any constant number from each number shown in the examples. For example, add 14 to each number, or subtract 1 from each number, etc.)

20. STEEPLE CHASE (Grades 4-8)

A. Preparation and Materials: Prepare a list of multiple-step problems suitable for children of your grade level to work mentally. You will need a stopwatch or a watch with a second hand to time responses.

Sample Problems:

1. $[(36 \div 2) \div 2] + 1 = ?$ (10)
2. $(45 + 7) + (3 + 15) = ?$ (70)
3. $[(99 - 33) \div 2] + 2 = ?$ (35)
4. $[(25 \times 2) \times 2] \div 4 = ?$ (25)
5. $[(18 \div 3) \times 2] \div 4 + 1 = ?$ (4)
6. $[(83 - 8) \div 3] + 11 = ?$ (36)
7. $[(17 + 5) \div 2] + 4 = ?$ (15)
8. $[(7 \times 3) + 4] - 6 = ?$ (19)
9. $[(5 \times 3) + 5] \div 4 = ?$ (5)
10. $[(18 + 2) - 3] + 4 = ?$ (21)

B. Introduction to the Class: Here is a game that will brush away your mental cobwebs. I will give a mathematics problem, I will pause 2 seconds. Be thinking of the answer during that pause. Then, I will call a girl's name. If she can answer within 2 more seconds, she wins a point for the girls.

For example, I might say $4 + 8, \div 2, \times 3, \div 9 = ?$ (Pause two seconds.) Mary? Yes, the answer is 2. So you would win one point for the girls. Next, I will give a new problem and call on a boy.

I will keep track of your scores on the board. Who do you think will get the most points, the girls or the boys?

*21. MATHEMATICS VOCABULARY
(Grades 4-8)

A. Preparation and Materials: Duplicate materials like that shown in the example making one copy for each child. In the example, the answers are circled for your convenience. Omit these circles from the duplicated sheet. Adapt the terms included to suit the ability level of your group. Children will need pencils.

Example:

1. Sum 46 + 37 (83)	2. Addend (46) +(37) 83	3. Difference 46 - 37 (9)
4. Subtrahend 46 -(37) 9	5. Minuend (46) - 37 9	6. Multiplicand (35) x 5 175
7. Product 35 x 5 (175)	8. Multiplier 35 x (5) 175	9. Exponent $3^{(3)} = 3 \times 3 \times 3 = 27$
10. Divisor 17 (3)⟌51	11. Quotient (17) 3⟌51	12. Dividend 17 3⟌(51)
13. Numerator (3) 8	14. Denominator 3 (8)	15. Proper Fraction (3/8) 4/3
16. Improper Fraction 3/8 (4/3)	17. Mixed Numeral 3/8 (1¼) 3	18. Univeral Set a. Doug & Don (b. All boys in town) c. Jack, Joe, Tom

—26—

B. Introduction to the Class: Look at the first square. What word is written in that square, Jerry? Yes, "Sum." Look at the problem written in that square. What part of that problem is the sum, Todd? Yes, 83. Circle the number 83.

In each of the other squares circle the part of the problem that illustrates the word given at the top of that square.

22. GUESSTIMATION (Grades 4-8)

A. Preparation and Materials: Prepare a collection of objects and questions as described in the sample questions. Children will need pencils and paper.

B. Introduction to the Class: Here is a game to test your ability to estimate weights and measurements. I'll show you an object and ask a question. You cannot measure, weigh or count what I show you. You must **estimate** each answer.

Please number your papers from 1 to 20. By each number write your estimate for that object.

(Show one object at a time, ask the question and allow children time to write their answers before proceeding to the next question. Keep track of the order in which objects are shown. When all have been shown, choose a child to weigh, measure or count each object and report the correct answer.)

Sample Questions:

1. Show a quantity of dried beans in a glass container.

QUESTION: How many beans are in this jar?

2. Show a full box of salt.

QUESTION: How much does this box of salt weigh?

3. Show a large dictionary or encyclopedia.

QUESTION: How many pages are in this book?

4. Show a ball of yarn.

QUESTION: How long is the strand of yarn in this ball?

5. Show a stick.

QUESTION: How many inches long is this stick?

6. Show a stack of paper.

QUESTION: How many sheets of paper are in this stack?

7. Show a basketball.

QUESTION: What is the circumference of this ball in inches?

8. Show a baseball.

QUESTION: What is the circumference of this ball in inches?

9. Show a book.

QUESTION: How much does this book weigh?

10. Show a book much larger or much smaller than the previous one.

QUESTION: How much does this book weigh?

11. Show a bundle of drinking straws enclosed in a rubber band.

QUESTION: How many straws are in this bundle?

12. Show a large bath towel.

QUESTION: What is the length of this towel in inches?

13. Show a glass container of water.

QUESTION: How many cups of water are in this jar?

14. Show a crayon.

QUESTION: How many of these crayons placed end to end would make a length of two feet?

15. Show a can or jar and a gallon jug.

QUESTION: How many times would I need to fill this container and pour it into the gallon jug before the jug was filled?

16. Point to a bookshelf in the classroom.

QUESTION: How many books are on that shelf?

17. QUESTION: What is the length of this classroom in feet?

18. Show a teaspoon full of dry rice in a glass.

QUESTION: How many grains of rice are in this glass?

19. Point out a child of average weight in the classroom.

QUESTION: How much does (name) weigh?

20. Point out a child of average height in the classroom.

QUESTION: How tall is (name)?

23. GRAPH DISPLAY (Grades 4-8)

A. Preparation and Materials: Have bulletin board space available. Title the display "Types of Graphs." You will need bulletin board pins, colored paper for mounting the graphs, etc.

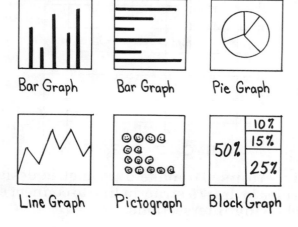

Bar Graph   Bar Graph   Pie Graph

Line Graph   Pictograph   Block Graph

B. Introduction to the Class: We have talked about the many types of graphs, and today, we are going to begin working on a bulletin board display of graphs.

When you find a graph in a magazine or newspaper, cut it out and bring it to school. You may mount the graph you bring, label its type and add it to the display. Let's see how many different types of graphs we can find.

** 24. GRAPH PLOTTING DESIGN (Grades 4-8)

A. Preparation and Materials: This activity is designed to give children practice in graph reading by developing skill in finding exact locations on a grid pattern.

Duplicate copies of the grid pattern and directions shown in the example, making one copy for each child. Children will need rulers and pencils, crayons or marking pens.

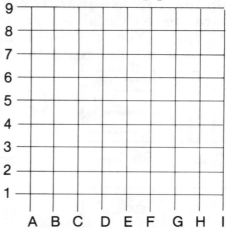

Directions: To make a geometric design use your ruler to draw a straight line to connect each of the following intersections:

*This activity is available in Challenge Volume I of the **Spice**™ Duplicating Masters.
This activity is available in Challenge Volume II of the **Spice™ Duplicating Masters.

1. 5-A and 1-E	10. 1-I and 5-I
2. 9-E and 5-I	11. 9-A and 1-I
3. 1-A and 1-E	12. 3-G and 7-G
4. 3-C and 3-G	13. 3-C and 7-C
5. 1-A and 9-I	14. 1-A and 9-A
6. 5-A and 9-E	15. 5-I and 9-I
7. 5-A and 5-I	16. 7-C and 7-G
8. 1-E and 5-I	17. 9-E and 9-I
9. 9-A and 9-E	18. 1-E and 1-I

B. Introduction to the Class: Each intersection on this grid can be described by giving the number of the horizontal line and the letter of the vertical line that cross at that point.

Can you find intersection 5-D? Can you find intersection 8-G? (Make sure each child understands how to find the exact location described by a number and letter pair before proceeding with this activity.)

Now, look at the first printed direction. Make a small dot at intersection 5-A and a dot at intersection 1-E. Use your ruler to help you draw a straight line to connect those two dots.

Complete the rest of the work in this same way. If you follow directions exactly, you will draw an interesting geometric design.

Child's Completed Work:

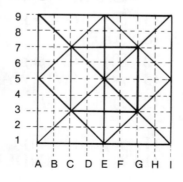

25. GRAPH SPELLING (Grades 4-8)

A. Preparation and Materials: This activity is designed to give children practice in graph reading by finding exact locations on a grid pattern.

Duplicate material like that shown in the example making one copy for each child. Children will need pencils.

Example:

5	U	V	W	X	Y
4	P	Q	R	S	T
3	K	L	M	N	O
2	F	G	H	I	J
1	A	B	C	D	E
	★	●	✔	□	?

Part I: Plot each symbol pair to spell a word.

1. (2 - ?)　　(5 - ★)　　(2 - ●)　　(2 - ●)
　 (3 - ●)　　(1 - ?)　　(JUGGLE)

2. (3 - ?)　　(4 - ★)　　(1 - ?)　　(4 - ✔)
　 (1 - ★)　　(OPERA)

3. (4 - □)　　(3 - □)　　(1 - ★)　　(3 - ★)
　 (1 - ?)　　(SNAKE)

4. (5 - ✔)　　(2 - □)　　(2 - ●)　　(5 - ✔)
　 (1 - ★)　　(3 - ✔)　　(WIGWAM)

5. (1 - ●)　　(3 - ?)　　(4 - ?)　　(2 - ✔)
　 (BOTH)

6. (1 - ✔)　　(1 - ★)　　(3 - □)　　(1 - □)
　 (3 - ●)　　(1 - ?)　　(CANDLE)

*This activity is available in Challenge Volume I of the Spice™ Duplicating Masters.
**This activity is available in Challenge Volume II of the Spice™ Duplicating Masters.

7. (2 - ★)   (3 - ?)   (5 - □)   (FOX)
8. (4 - ●)   (5 - ★)   (2 - □)   (5 - ●)
   (1 - ?)   (4 - ✔)   (QUIVER)
9. (1 - ●)   (1 - ★)   (4 - ✔)   (2 - ●)
   (1 - ★)   (2 - □)   (3 - □)   (BARGAIN)
10. (4 - ?)   (3 - ?)   (4 - ?)   (1 - ★)
    (3 - ●)   (TOTAL)
11. (1 - □)   (1 - ★)   (3 - □)   (1 - ✔)
    (1 - ?)   (DANCE)
12. (4 - ✔)   (1 - ★)   (1 - ●)   (1 - ●)
    (2 - □)   (4 - ?)   (RABBIT)
13. (4 - ?)   (4 - ✔)   (5 - ★)   (3 - □)
    (3 - ★)   (TRUNK)
14. (4 - □)   (4 - ?)   (3 - ?)   (4 - ✔)
    (3 - ✔)   (STORM)
15. (4 - ★)   (5 - ★)   (4 - ✔)   (4 - ★)
    (3 - ●)   (1 - ?)   (PURPLE)

Part II: Write the symbol pairs to plot the spelling of these words.

1. STEM   (4 - □) (4 - ?) (1 - ?) (3 - ✔)
2. ALSO   (1 - ★) (3 - ●) (4 - □) (3 - ?)
3. FORMED   (2 - ★) (3 - ?) (4 - ✔) (3 - ✔)
              (1 - ?) (1 - □)
4. BUSY   (1 - ●) (5 - ★) (4 - □) (5 - ?)
5. LIGHT   (3 - ●) (2 - □) (2 - ●) (2 - ✔)
              (4 - ?)
6. KNOW   (3 - ★) (3 - □) (3 - ?) (5 - ✔)
7. ACTIVE   (1 - ★) (1 - ✔) (4 - ?) (2 - □)
              (5 - ●) (1 - ?)
8. PACK   (4 - ★) (1 - ★) (1 - ✔) (3 - ★)

9.	EXIT	(1 - ?)	(5 - □)	(2 - □)	(4 - ?)
10.	GRAPH	(2 - ●)	(4 - ✓)	(1 - ★)	(4 - ★)
		(2 - ✓)			
11.	BRIDGE	(1 - ●)	(4 - ✓)	(2 - □)	(1 - □)
		(2 - ●)	(1 - ?)		
12.	PENCIL	(4 - ★)	(1 - ?)	(3 - □)	(1 - ✓)
		(2 - □)	(3 - ●)		
13.	FIRST	(2 - ★)	(2 - □)	(4 - ✓)	(4 - □)
		(4 - ?)			
14.	MUSIC	(3 - ✓)	(5 - ★)	(4 - □)	(2 - □)
		(1 - ✓)			
15.	STRING	(4 - □)	(4 - ?)	(4 - ✓)	(2 - □)
		(3 - □)	(2 - ●)		

B. Introduction to the Class: Each letter in this diagram is placed on the intersection of two lines. The exact location of each intersection can be described by naming the number of that line across and the symbol for that line up.

For example, what letter is at the intersection of lines 3 - ✓, Julie? Yes, M. What letter is at the intersection of lines 4 - □, John? Yes, S. (Continue in this way until you are sure all children understand how to locate each intersection.)

In the first section of this activity, write the letter you find at the intersection described by each pair of numbers and symbols. The letters described in each line will spell a common word.

In the second part of this activity, use number and symbol pairs to describe the exact location of each letter in each word given.

If you and a friend each make a copy of a graph of this kind using symbols you create yourselves, you can write messages in code.

—34—

*26. SCRAMBLE (Grades 5-6)

A. Preparation and Materials: On the board draw and fill in a chart like the example shown. Children will need paper and pencils.

Example:

Whole Numbers	Mixed Numbers	Improper Fractions	Proper Fractions
$1\frac{3}{8}$	$2\frac{2}{3}$	8	24
7	$\frac{6}{3}$	3	$\frac{36}{37}$
$\frac{9}{4}$	$\frac{11}{10}$	$\frac{11}{17}$	$8\frac{1}{7}$
$\frac{4}{5}$	$\frac{5}{7}$	$\frac{14}{9}$	$1\frac{1}{2}$
$5\frac{3}{4}$	6	$\frac{7}{8}$	31
$\frac{5}{4}$	$\frac{3}{8}$	$9\frac{2}{3}$	$\frac{9}{2}$

B. Introduction to the Class: On your paper draw and label the four columns as you see them on the board. I have written many numbers on the board, but they are in no special order. I would like you to separate these numbers on your paper so only the whole numbers will be written in the WHOLE NUMBERS column, only mixed numbers in the MIXED NUMBERS column, etc.

Read the first number. Decide in which column that number belongs. Write it in the proper column. Then read the next number and place it where it belongs, and so on.

*This activity is available in Challenge Volume I of the Spice™ Duplicating Masters.

Example of Child's Completed Paper:

Whole Numbers	Mixed Numbers	Improper Fractions	Proper Fractions
7	$1\frac{3}{8}$	$\frac{9}{4}$	$\frac{4}{5}$
6	$5\frac{3}{4}$	$\frac{5}{4}$	$\frac{36}{37}$
8	$2\frac{2}{3}$	$\frac{6}{8}$	$\frac{5}{7}$
3	$9\frac{2}{3}$	$\frac{14}{9}$	$\frac{3}{8}$
24	$8\frac{1}{7}$	$\frac{3}{2}$	$\frac{11}{17}$
31	$1\frac{1}{2}$	$\frac{11}{10}$	$\frac{7}{8}$

27. FRACTION RACE (Grades 5-6)

A. Preparation and Materials: Make 20 flashcards, each showing a mixed number and its improper fraction equivalent. Omit the numerator of each improper fraction.

Example:

$$3\frac{1}{4} = \frac{}{4} \qquad 2\frac{2}{3} = \frac{}{3} \qquad 1\frac{7}{8} = \frac{}{8}$$

$$6\frac{2}{5} = \frac{}{5} \qquad 2\frac{1}{7} = \frac{}{7} \qquad 4\frac{5}{6} = \frac{}{6}$$

Divide the class into two equally matched teams. Each child will remain at his own desk, and will need a pencil and paper.

B. Introduction to the Class: Number your papers from 1 to 20. I will hold up a card. You are to write the missing number in the equation I show you.

After I have shown all the cards, we will correct your papers in class. If you have a perfect score, you will earn two points for your team. If you miss 1 or 2 problems, you will earn one point. If you miss three or more, you will earn no score. The team with the highest total score wins.

Ready? Here is the first equation.

28. FRACTION PROOF (Grades 5-6)

A. Preparation and Materials: Write on the board problems in subtraction of fractions from the number 1.

Examples:

1 - 3/5 = 2/5,   1 - 3/4 = 1/4,   1 - 3/8 = 5/8

Each child will need a sheet of ¼ inch graph paper, scissors, colored pencils or crayons, paste and a sheet of construction paper.

B. Introduction to the Class: On the board are examples of subtracting a fraction from the whole number 1. The answers are correct. We are going to make graph paper illustrations to prove these answers are correct.

First, cut a section which will represent the whole number 1. Divide it into as many parts as the denominator of the fraction indicates. Cut away the number of parts needed to illustrate each example. The number of parts left should be the same as the answer given on the board.

Mount each completed illustration on construction paper. Under it write the problem which it illustrates. Work neatly for I will select some of the neater papers for bulletin board display.

Example:

$$1 - \tfrac{3}{5} = \tfrac{2}{5} \qquad\qquad 1 - \tfrac{3}{4} = \tfrac{1}{4}$$

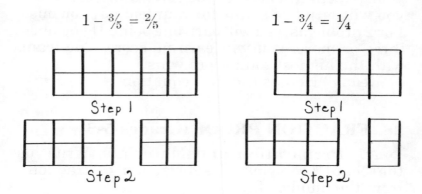

Step 1   Step 1

Step 2   Step 2

29. REDUCTION RELAY (Grades 5-6)

A. Preparation and Materials: Write on the board a number of fractions which are not reduced to their lowest terms. Divide the class into two teams.

B. Introduction to the Class: I will point to a fraction on the board. If the first person on Team A can correctly reduce this fraction to its lowest terms, he will win a point for his team. Then, I will call on the first person from Team B, and so on. At the end of the playing time the team with the highest score wins.

C. Variation: Write on the board a list of fractions which are not reduced to their lowest terms. Divide the class into two teams. Ask all children to write each fraction reduced to its lowest terms. When papers are corrected, the team with the highest total of correct answers wins. This variation allows all children to participate simultaneously and thus provides more individual drill.

30. TAKE THREE CARDS (Grades 5-8)

A. Preparation and Materials: Cut 60 two-inch squares of paper. Write the numbers from 1 to 20 writing one number on each square. Make three duplicates of each number. Put all the squares in a box.

B. Introduction to the Class: Today, we will work together deciding upon common denominators. Each of you may have a turn taking three cards from this box. You may imagine that the numbers are denominators in a fraction problem.

Show the class the three numbers you draw and then tell the lowest common denominator you would use to work the problem.

Suppose you drew:

You would select 40 as the lowest common denominator.

C. Variation: Place a number of squares, each containing a fraction or mixed number, in a box. Have children draw three squares, and use the numbers for an addition problem which they may work at the board.

**31. EVERYDAY FRACTIONS (Grades 5-8)

A. Preparation and Materials: Prepare a list of questions like those in the example. The questions can be given orally or written on the board for children's reference.

This activity is available in Challenge Volume II of the **Spice™ Duplicating Masters.

Example:

1. 9¢ are what part of a dollar? _(9/100)_
2. 3 hours are what part of a day? _(3/24)_
3. 5 eggs are what part of a dozen? _(5/12)_
4. 17 inches are what part of a yard? _(17/36)_
5. 5 ounces are what part of a pound? _(5/16)_
6. 31 days are what part of a year? _(31/365)_
7. 7 inches are what part of a foot? _(7/12)_
8. 2 days are what part of a week? _(2/7)_
9. 1 quart is what part of a gallon? _(1/4)_
10. 63 feet are what part of a mile? _(63/5,280)_
11. 3 pecks are what part of a bushel? _(3/4)_
12. 1 pint is what part of a quart? _(1/2)_

B. Introduction to the Class: In a fraction the denominator tells how many pieces make up the whole group or object. The numerator tells how many of those pieces we have.

Look at the first example on the board. We are discussing cents in a dollar. How many cents make up the whole dollar, Bill? Yes, 100. So 100 will be the denominator of the fraction. How many of those 100 cents do we have, Jill? Yes, 9. So 9 will be the numerator of the fraction. What is the whole fraction, Martha? Yes, 9/100.

Complete the remainder of the problems in this same way.

*32. ROUNDING OFF (Grades 5-8)

A. Preparation and Materials: Place on the board three rows of addition problems suitable for the ability level of your group. Each child will need a pencil and paper.

Example:

Row 1 Tens	Row 2 Hundreds	Row 3 Thousands
245 _____	3457 _____	10,534 _____
349 _____	599 _____	9,999 _____
399 _____	2492 _____	15,659 _____
265 _____	4224 _____	11,431 _____
439 _____	1549 _____	29,569 _____
99 _____	1299 _____	13,321 _____

B. Introduction to the Class: On the board are three rows of addition problems. I would like you to round off the addends in row one to the nearest ten, in row two to the nearest hundred and in row three to the nearest thousand. Then work each problem in the usual way.

Example Rounded Off to the Nearest Hundred:

4751 4800
3643 3600
6444 6400
8988 9000

*This activity is available in Challenge Volume I of the **Spice**™ Duplicating Masters.

*33. PERIMETERS (Grades 5-8)

A. Preparation and Materials: Draw the example shown on the board, or duplicate the work making one copy for each child.

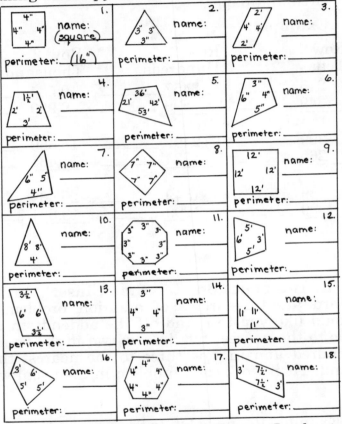

B. Introduction to the Class: Look at the first section. What is the name of the shape in that section, Alice? Yes, a square. How did you know it was a square? (All four sides are of equal length.)

What is the perimeter of this square, Don? Yes, 16 inches. How do you know? (To find the perimeter add the length of all the sides.)

*This activity is available in Challenge Volume I of the **Spice**™ Duplicating Masters.

By the number 1 on your paper write the name of the shape shown in that section, and the perimeter of that shape. Complete the other sections in this same way.

**34. DECIMAL PLACE VALUE CHART (Grades 5-8)

A. Preparation and Materials: Place an example of the chart on the blackboard. Children will need paper and pencils.

Example:

	Hundreds	Tens	Ones	Tenths	Hundredths	Thousandths
1.	4	2	8			
2.		4	2	8		
3.			4	2	8	
4.						
5.						
6.						
7.						
8.						
9.						
10.						
11.						
12.						

B. Introduction to the Class: Will you each please make a chart like the one you see on the board? Now, see if you can place these numbers correctly on your charts:

(Write these numbers on the board.)

1. 428
2. 42.8
3. 4.28
4. 4.289
5. 12/10
6. 12322/100
7. 203.06
8. 74/1000
9. 6.7
10. 4.003
11. 43/100
12. 849/100

—43—

C. Variation: Children may make up a list of ten numbers and exchange with each other. The child who composed the list may correct his friend's answers, with the teacher arbitrating when necessary.

**35. SQUARES OF 100 (Grades 6-8)

A. Preparation and Materials: Each child will need a sheet of graph paper, colored pencils, scissors, paste and a sheet of construction paper.

B. Introduction to the Class: On your graph paper count ten squares down and ten squares across. Mark around this large square with your pencil and cut the square out. How many small squares does this large square contain, Bob? Yes, 100.

Color 22 squares red. What part of the whole did you color red? (22/100) Can you write that fraction as a decimal? (.22)

Today, you are going to make squares showing hundredths. You will need to cut one large square of graph paper showing ten squares down and ten squares across to illustrate each of the examples I will now put on the board.

Example:

1. .06	4. .28	7. .08
2. .42	5. .33	8. .75
3. .19	6. .98	9. .64

After you have cut and colored your square, mount them all on one sheet of construction paper. Below each illustration label the decimal fraction shown by that square. Work as neatly as you can. I will select some of the neater papers to display on the bulletin board.

—44—

Example:

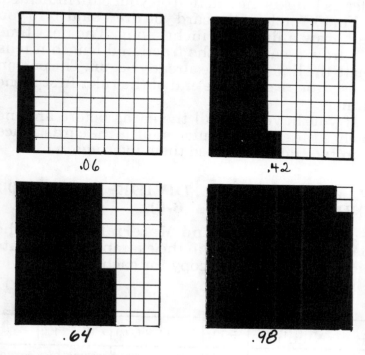

.06   .42

.64   .98

36. RULER RACE (Grades 6-8)

A. Preparation and Materials: Cut 20 small squares of paper. On each square write a fraction whose denominator is 2, 4, 8, or 16. Provide a ruler scaled to show sixteenths of an inch, a sheet of paper and colored pencils.

B. Introduction to the Class: Two to four players may play this game of Ruler Race. Place the edge of the ruler against the edge of the paper. Place the cards in a stack face down in the center of the table. Each player will need a pencil of a different color to mark his score.

Take turns drawing one card from the stack. Use your colored pencil to draw a line above the ruler as long as the fraction on your card indicates.

For example, if the card you drew said 5/16 you may draw a line 5/16 inches long. The next time you draw a card add the fractional length it tells you to the line you have already begun. The person whose line reaches the end of the ruler first is the winner.

If you have used all the cards before anyone reaches the end of the ruler, shuffle the cards, place them face down and use them over again.

**37. MULTIPLY OR DIVIDE BY 10, 100 AND 1,000 (Grades 6-8)

A. Preparation and Materials: Put on the board the chart shown in the example or duplicate the work making one copy for each child.

Example:

Base Number	÷ 10	÷ 100	÷ 1,000	x 10	x 100	x 1,000
1. 36.4	(3.64)	(.364)	(.0364)	(364)	(3,640)	(36,400)
2. 7.423						
3. 135.08						
4. 62.1						
5. .03						
6. 421.6						
7. 3.97						
8. .2						
9. .02						
10. .002						

**This activity is available in Challenge Volume II of the Spice™ Duplicating Masters

B. Introduction to the Class: We have talked about moving the decimal point to the right or left to multiply or divide a number by 10, 100, or 1,000. (Review moving the decimal point to the left or right to divide or multiply by 10, 100 and 1,000.) Now, we'll see how well you remember these rules.

Look at the first number, 36.4. Where should the decimal point be placed if you divide that number by 10, Jason? (3.64) Where would the decimal point be placed if you divide that number by 100, Kathy? (.364) Continue this way to complete the first row of the chart.

Now, I would like you to finish the other rows of the chart in this same way.

C. Variation: Make the same kind of chart, but in each row give a number in any column of the chart. Children fill in the remainder of the row from the one number given.

Example:

Base Number	÷ 10	÷ 100	÷ 1,000	x 10	x 100	x 1,000
1.	4.8					
2.				22.5		
3.			.005			
4.						42,195
5. .36						
6.		.204				
7.	37.3					
8. 5.23						
9.						1,018
10.			.7374			

** 38. RING AROUND (Grades 6-8)

A. Preparation and Materials: Write on the board several rows of fractions with unlike denominators. Children will need paper, pencils and crayons.

Example:

1. $\left(\frac{2}{3}\right)$ ← red	$\frac{1}{2}$	$\frac{1}{4}$	$\frac{3}{8}$	$\left(\frac{1}{6}\right)$ ← blue	$\frac{1}{3}$
2. $\frac{1}{2}$	$\frac{3}{16}$	$\frac{5}{8}$	$\frac{1}{4}$	$\frac{3}{4}$	$\frac{3}{8}$
3. $\frac{2}{5}$	$\frac{14}{15}$	$\frac{2}{3}$	$\frac{5}{6}$	$\frac{1}{2}$	$\frac{1}{6}$
4. $\frac{5}{12}$	$\frac{5}{6}$	$\frac{1}{3}$	$\frac{13}{18}$	$\frac{2}{3}$	$\frac{3}{4}$
5. $\frac{11}{14}$	$\frac{5}{7}$	$\frac{1}{2}$	$\frac{2}{7}$	$\frac{5}{14}$	$\frac{1}{14}$
6. $\frac{1}{2}$	$\frac{5}{8}$	$\frac{11}{12}$	$\frac{1}{12}$	$\frac{2}{3}$	$\frac{3}{4}$
7. $\frac{5}{9}$	$\frac{13}{18}$	$\frac{2}{3}$	$\frac{1}{3}$	$\frac{5}{18}$	$\frac{2}{9}$
8. $\frac{11}{15}$	$\frac{4}{5}$	$\frac{1}{3}$	$\frac{2}{3}$	$\frac{2}{15}$	$\frac{1}{5}$
9. $\frac{4}{7}$	$\frac{19}{28}$	$\frac{3}{4}$	$\frac{11}{14}$	$\frac{6}{7}$	$\frac{23}{28}$
10. $\frac{3}{5}$	$\frac{19}{25}$	$\frac{1}{3}$	$\frac{14}{15}$	$\frac{4}{5}$	$\frac{14}{25}$

B. Introduction to the Class: Please copy each row of fractions you see on the board. In each row please draw a red ring around the largest quantity and a blue ring around the smallest quantity. You can convert all fractions in a row to like denominators to determine their relative value.

39. FRACTION DOMINOES (Grades 6-8)

A. Preparation and Materials: Cut "dominoes" from tagboard, using pieces 2 x 4 inches. On one side write a common fraction, and on the other side a decimal fraction. The items on the two sides of a single domino will not be equal. Make four copies of each domino.

Example:

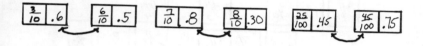

B. Introduction to the Class: Here is a fraction game that is played like regular dominoes. Place all the dominoes face down in the center of the table. Four players may draw four cards each.

The first player may place any one of his dominoes in the center, and succeeding players may match the decimal fraction or common fraction with its equivalent (demonstrate).

Example:

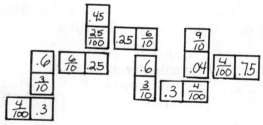

A person may play only one domino per turn. If he cannot match any one of his dominoes with one already played on the board, he draws another domino from the center and play passes on. The first player to use all his dominoes is the winner.

C. Variation: Use improper fractions and their reduced forms on the dominoes. Play in the same manner.

Example:

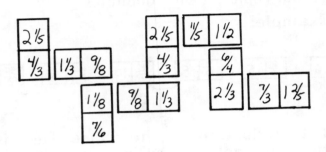

40. CHAIN REACTION (Grades 6-8)

A. Preparation and Materials: None.

B. Introduction to the Class: I will give two numbers to the first person in row 1 and tell him what to do with them. He will give the answer, then tell the person behind him how to proceed from there. The game will go like this:

Teacher: "2½ and 2. Multiply."
Child 1: "The answer is 5. Add ¾."
Child 2: "The answer is 5¾. Subtract 2½."
Child 3: "The answer is 3¼. Divide by ⅛."

We will continue to work in this way until someone gives a wrong answer. Then, the game starts all over again. Let's see how far we can go around the room before someone makes a mistake.

(If the numbers being used become too complicated, either step into the game and give directions which will get the number back to workable dimensions or stop the game and begin with a new number.)

41. EQUIVALENT WHEEL (Grades 6-8)

A. Preparation and Materials: Cut a tagboard circle at least 12 inches in diameter. Around the edge write a variety of decimal fractions. To the center of the circle, attach a tagboard spinner with a paper fastener, making sure the hand can spin freely. Divide the class into two teams.

Example:

B. Introduction to the Class: The first person from Team A may come up to the wheel and spin the hand. When the hand stops, he will look at the number to which it points. He will tell us three things:

1. The decimal fraction itself.
2. The common fraction equivalent.
3. Its equivalent in percent.

If he answers all three correctly, he wins a point for his team. Then the first person from Team B may use the wheel. At the end of the playing time the team with the highest score wins.

C. Variations: Place improper fractions on the wheel, and each player tells the reduced form of the fraction. Or, using a 24 inch wheel place problems in addition, subtraction, etc., using either whole numbers or fractions, around the rim. Each player must correctly solve the problem to which the hand points to win a score for his team. Or, you could place Roman numerals around the rim. Children must tell the Arabic equivalent.

**42. COMMON DENOMINATOR (Grades 6-8)

A. Preparation and Materials: Cut 20 tagboard flashcards. On each card write two or three unlike fractions. Children will need pencils and paper.

Example:

$$\boxed{\dfrac{1}{2} \quad \dfrac{1}{4}} \quad \boxed{\dfrac{3}{5} \quad \dfrac{1}{2}} \quad \boxed{\dfrac{2}{3} \ \dfrac{1}{2} \ \dfrac{3}{4}} \quad \boxed{\dfrac{5}{6} \ \dfrac{1}{2} \ \dfrac{2}{3}} \quad \boxed{\dfrac{1}{5} \ \dfrac{1}{3} \ \dfrac{3}{10}} \quad \boxed{\dfrac{2}{5} \ \dfrac{1}{7} \ \dfrac{7}{10}}$$

B. Introduction to the Class: I will show you a flashcard on which will be written two or three unlike fractions. You must think of the lowest possible common denominator for these fractions, and write that number on your paper.

First, number your papers from one to twenty. Beside the number "1," write the lowest common denominator for the fractions I will show you now. (Proceed in this way until all the cards are shown. Jot down the answer as each card is shown and use this list for correcting the children's completed work.)

C. Variations: Divide the class into two equally matched teams. Show cards to the teams in alternate order. A child scores one point for his team for each correct answer.

**This activity is available in Challenge Volume II of the Spice™ Duplicating Masters.

Or, by writing the correct answer on the back of each card, the game can be adapted for independent use. One child may show his friend the cards. The friend will give the lowest common denominator. The child holding the cards can check each answer given against the answer written on the back of the cards.

**43. A PUZZLE IN TENTHS (Grades 6-8)

A. Preparation and Materials: Place on the board problems involving the use of tenths. Cover these problems until the time when children actually begin their written work. Each child will need a pencil and paper.

Example:

1. $10 \times .1 = ?$
2. $10 \times 1/10 = ?$
3. $10 \times 1 = ?$
4. Write as decimals: 3/10, 4 and 4/10, 22 and 2/10, 7/10.
5. Count by tenths from 3 to 4.
6. Arrange these numbers by value, putting the lowest first and the highest last: 2/10, 4/10, 2 and 6/10, 3/10, 4 and 1/10, 2 and 5/10.
7. How many tenths in one inch?
8. Fill in the blanks: .0 _ _ _ .4 _ _ _ .8
9. The odometer of a car showed 212.3 at the beginning of a trip and 317.8 at the end. How many miles had the car traveled on this trip?
10. Write these decimals as fractions reduced to their lowest terms: .5, .8, .7, .4, .3.
11. Add together 3.2, .7, 2.6, 1.4, .5.
12. In what order are these numbers: .8, .5, .4, .9, .1, .7, .6, .3 and .2? (Alphabetical order according to the written word for each numeral.)

—53—

44. ROMAN RACE (Grades 6-8)

A. Preparation and Materials: Children will need paper and pencils.

B. Introduction to the Class: I will give you ten minutes to write the numbers from 1 to 100. All those with perfect papers may be dismissed first for lunch. Doesn't that sound easy? There's only one minor rule to this game. Please count in **Roman numerals.** Go! (Check papers in class after the 10 minute period is up.)

45. COMMON SENSE WITH DECIMALS (Grades 6-8)

A. Preparation and Materials: Prepare a list of questions like those shown in the example.

B. Introduction to the Class: Have you ever missed a mathematics problem because you put a decimal point in the wrong place? A misplaced decimal point can change a quantity of 5 into a quantity of 50, or 500 or even 5,000 which is a gigantic change indeed.

Before you work a problem, think of the **approximate** answer. After you have worked the problem, check to see if your answer is sensible in terms of that approximation.

For example, if you are adding 5.3 and 6.7, will you answer to 12 or 120, Saul? Yes, 12. How can you tell? (The whole numbers 5 and 6 almost add up to 12, and the decimal fractions make up the rest. The whole numbers 5 and 6 will total nowhere near 120, so 120 would be totally unsensible.)

If you took a dollar bill to the store to buy an item that cost 96¢, would your change be 4¢ or $1.04, Patty? Yes, 4¢. How can you tell? (96¢ is almost a whole dollar so the change would be a very small amount. Also, $1.04 is more than the amount you had in the beginning so would be a totally unsensible answer.)

I will give you some more problems of this kind to answer orally. After you choose the sensible answer, please tell me why that choice is sensible and why the other choice is totally unsensible.

Example:

1. A farmer sold 6 dozen eggs at 60¢ a dozen. Did he collect $3.60 or $36.00?

2. Three boys each bought 3 candy bars for 10¢ each. Did all the candy bars cost $9.00 or 90¢?

3. A square is 4 inches long on each side. Is the perimeter of that square 1.6 inches or 16 inches?

4. John ran a mile a day to keep in shape for sports. At the end of the week had he run 7 miles or 70 miles?

5. A man died and left $2,000 to each of his three daughters. Did the girls together receive $60,000 or $6,000?

6. Mary read 37 pages of a book on Saturday and 42 on Sunday. On these two days did she read 79 pages or 790 pages?

7. A boy saved 10¢ a day for two weeks. Did he save $1.40 or $14.00?

8. Tom cut 4 apples in half to share with his friends. Did he cut the apples into .8 pieces or 8 pieces?

9. A garden is 40 feet long and 20 feet wide. Is the area of this garden 800 square feet or 8,000 square feet?

10. On three successive math tests Lisa scored 85%, 92% and 78%. Was her average score 8.5% or 85%?

Remember this activity the next time you are working problems with decimal points. If your answer does not make sense in relation to the question asked, check the placement of your decimal point. It can make a world of difference.

**46. GRAPH CODE MESSAGE (Grades 6-8)

A. Preparation and Materials: Children will need graph paper and pencils or marking pens. Directions can be written on the board, but it is easier on the eyes if directions are duplicated so each child has one of his/her own.

1. (9N-6W)	(5N-6W)	20. (8S-10W)	(8S-8W)
2. (9N-4W)	(7N-6W)	21. (4S-7W)	(8S-7W)
3. (7N-6W)	(5N-4W)	22. (4S-7W)	(8S-6W)
4. (9N-3W)	(5N-3W)	23. (4S-5W)	(8S-6W)
5. (9N-3W)	(9N-1W)	24. (4S-5W)	(8S-5W)
6. (7N-3W)	(7N-2W)	25. (4S-4W)	(8S-4W)
7. (5N-3W)	(5N-1W)	26. (4S-3W)	(8S-3W)
8. (9N-1E)	(5N-1E)	27. (8S-3W)	(8S-1W)
9. (9N-1E)	(9N-3E)	28. (4S-1E)	(8S-1E)
10. (7N-1E)	(7N-2E)	29. (4S-2E)	(8S-2E)
11. (5N-1E)	(5N-3E)	30. (4S-2E)	(8S-4E)
12. (9N-4E)	(5N-4E)	31. (4S-4E)	(8S-4E)
13. (9N-4E)	(9N-6E)	32. (4S-5E)	(8S-5E)
14. (7N-4E)	(7N-6E)	33. (4S-5E)	(4S-7E)
15. (9N-6E)	(7N-6E)	34. (8S-5E)	(8S-7E)
16. (4S-10W)	(4S-8W)	35. (6S-7E)	(8S-7E)
17. (4S-10W)	(6S-10W)	36. (6S-6E)	(6S-7E)
18. (6S-10W)	(6S-8W)	37. (3S-9E)	(7S-9E)
19. (6S-8W)	(8S-8W)	38. Put a big dot on (8S-9E)	

B. Introduction to the Class: Today, we are going to plot and connect locations on a 4-way grid pattern to discover a secret message.

**This activity is available in Challenge Volume II of the Spice™ Duplicating Masters.

First, we'll use graph paper to label the intersection of lines. Put a dot on an intersection of lines right in the middle of your graph paper. Guess where the middle is as you don't need to have the exact center.

Counting that dot as zero, draw a vertical line 10 squares up and 10 squares down from that dot. (Illustrate on the board, putting in the N-S line in the diagram.) Put N for north at the top of that line and S for south at the bottom. Mark a dot and number every intersection from 1 to 10 above and below that center dot. (Illustrate on the board.)

Now, draw a horizontal line going 10 squares to the right and 10 squares to the left of that center dot. Put W for west at the left of that line and E for east at the right. Put a dot at each intersection point and label the dots from 1 to 10 in each direction from the center. (Illustrate the E-W line on the board.)

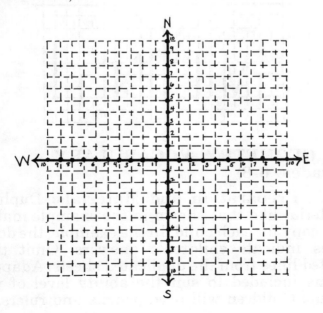

Now, you are ready to start plotting number locations to discover the secret message. Look at the first set of directions, (9N-6W) (5N-6W). On your graph paper find the intersection of lines 9N and 6W. Put a dot on that intersection. Find the intersection of lines 5N and 6W. Put a dot on that intersection. Draw a straight line to connect those two dots.

Put a dot on each of the other pairs of intersections described. Connect these pairs of dots and you will discover the secret message.

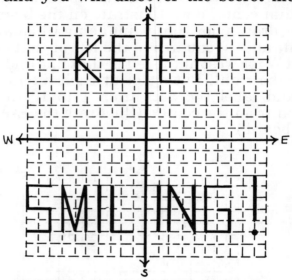

**47. GEOMETRY VOCABULARY (Grades 6-8)

A. Preparation and Materials: Duplicate material like that shown in the example making one copy for each child. In the example, the dotted lines indicate children's answers. Omit these dotted lines from the duplicated sheets. Adapt the terms included to suit the ability level of your group. Children will need pencils and rulers.

**This activity is available in Challenge Volume II of the Spice™ Duplicating Masters.

1. Line segment PQ P ●- - - - -● Q	2. A closed curve	3. A line passing through point R - - -●- - - R
4. Three parallel lines	5. A right angle	6. Two lines intersecting at point S
7. Line segments AB and BC A ●————● B ●C	8. An arc	9. A rectangle
10. A trapezoid	11. Two concentric circles	12. A line congruent with AB. A ● B - - - - - - -
13. A line perpendicular to EF E ●- - -┆- - -● F	14. Line segments AB, BC and CD C ●←B - - -→● D	15. A square figure
16. A line tangent to the circle	17. A ray passing through points M & N. M ●- - -●- - -→ N	18. Bisect the circle

B. Introduction to the Class: Read the description in each square. Draw the figure described.

**48. EQUIVALENT VALUE (Grades 7-8)

A. Preparation and Materials: Write the example on the board. Children will need pencils and paper.

B. Introduction to the Class: Look at the first number in column 1. It is 9/8. Now, look in column 2. Can you see a number in that column that has the same value as 9/8, Paul? Yes, 1 1/8. Do you see a number in column 3 that also has the same value, Randy? Yes, 1.125.

**This activity is available in Challenge Volume II of the Spice™ Duplicating Masters.

Example:

1.	9/8	1.	80%	1.	9/12
2.	4/5	2.	1.25	2.	66 2/3%
3.	2/3	3.	1⅛	3.	15/40
4.	1½	4.	75%	4.	90%
5.	1 2/5	5.	1.33	5.	9/6
6.	3/4	6.	.66	6.	1.4
7.	1¼	7.	7/5	7.	.80
8.	3/8	8.	1.5	8.	133 1/3%
9.	1 1/3	9.	.9	9.	1.125
10.	9/10	10.	.375	10.	125%

By the number 1 on your paper write 9/8. Beside that write the number from column 2 and the number from column 3 that have the same value as 9/8. Complete the rest of the work in this same way.

CHAPTER II:
"Games for One or Two"

Some of these games are centuries old and others are quite new. Many were created by mathematicians for their own personal leisure-time fun. All are thoroughly enjoyable to the child with a creative and analytical mind.

1. ODD OR EVEN (Grades 4-6)

A. Preparation and Materials: Provide a box of dried beans, corn, or any other small objects.

B. Introduction to the Class: Two players may try their luck in this Odd Or Even game. Each player starts the game with 20 beans. The first player puts some beans in his hand and holds them so the other player cannot see. The second player guesses if the number of hidden beans is Odd Or Even.

If his guess is correct, he wins the beans, If he is not correct, he must give that number of his own beans to his opponent.

The second player then hides a quantity of beans in his hand, and so on.

The winner is the player who collects all of his opponent's beans, or who has the most beans at the end of the playing time.

2. CIRCLE DRAW (Grades 4-8)

A. Preparation and Materials: A quantity of counters, such as checkers, squares of paper, kernels of corn, etc.

B. Introduction to the Class: Here is a game two of you may play in your free time. Arrange any number of counters in a circle. (Place the example diagram on the board.)

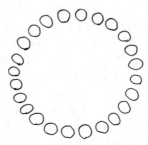

The two players take turns removing either one counter or two counters from the circle. If you take two counters, they must be ones that are side by side, with no other counter or empty space between them. The winner is the person who takes the last counter from the circle.

Let's try one game just to make sure everyone understands. Brad, will you and Tom come to the board? The two of you may take turns erasing either one circle or two circles, and let's see who wins by erasing the very last circle.

3. ELIMINATE (Grades 4-8)

A. Preparation and Materials: A quantity of markers, such as stones, beads, squares of paper, etc.

B. Introduction to the Class: When you have free time, you may enjoy playing the game of Eliminate with a friend. To play this game arrange any number of stones, beads, squares of paper, etc., into three rows.

The two players take turns in moving. A move consists of taking away any number of the articles from **one row only.** You may take any row, and you must always take at least one object when it is your turn. But you may take two or three objects, or even a whole row, if you wish to. The player who must take the last object is the loser.

This game may appear to be very simple, but once you've tried it you will realize it takes a great deal of strategy and cunning.

(Demonstrate this game so you are sure all the children understand. To demonstrate, you could draw three rows of circles on the board, and call on two children to play by erasing circles.)

4. ODD DRAW (Grades 4-8)

A. Preparation and Materials: You will need a quantity of pebbles, dried beans, paper squares or any other objects which can be used as counters.

B. Introduction to the Class: This is a game for two players. Place any number of counters in a pile on the table. Let's use 45 counters for this practice game. Next, we must decide the maximum number of counters a player may take on his turn. For this game we'll use a maximum of four.

Players alternately draw 1, 2, 3, or 4 counters from the pile and add them to their own piles of counters. After all counters have been drawn, the winner is the person who has drawn an odd number of counters. You can see you will need to keep a running total of the counters you have drawn, and plan a strategy for keeping that total an odd number.

Try it with different quantities of counters in the original pile, and with different maximums to be drawn on each turn. Just remember the original pile must contain an odd number of counters.

5. CAPTURE (Grades 4-8)

A. Preparation and Materials: You will need an egg carton containing two rows of 6 cups each, and 48 counters (beans, pebbles, beads, etc.).

B. Introduction to the Class: Here is a game two people can play in their free time. It is called Capture. This game is very popular on the Gold Coast in Africa.

Each player has one lengthwise half, or 6 cups, of the egg carton as his "side." Each cup contains 4 beans at the beginning of the game.

The players take turns moving. To move, you take all the beans from any one of the cups on your side of the egg carton, and moving clockwise around the egg carton, drop one bean in each cup as you pass over it.

The object of the game is to make Captures. You can make a Capture when your last bean goes on your opponent's side of the egg carton, **and** if that bean raises the total number of beans in that cup to 2 or 3. When this happens, you Capture all the beans in that cup and may set them aside as your winnings.

Also, after you make a Capture, you may look in the cup just before the one in which you made your Capture. If that cup now contains 2 or 3 beans, you may Capture those, also. You may continue capturing in this way, backwards around the egg carton, until you come to a cup that has less than 2 or more than 3 beans in it (or, until you reach your own side of the egg carton, as you cannot Capture on your own side).

The game ends when there are no more beans left in the egg carton, or when a player has no beans left on his side of the egg carton to move. In this case the opponent adds all the beans on his own side of the board to his own winnings, or there are so few beans left that it looks impossible to make any more Captures.

At the end of the game, the player who has captured the most beans is the winner.

Sample Plays:

1. Opening play: Player B takes all his beans from cup 5 and distributes them clockwise around the board.

Before Play

After Play

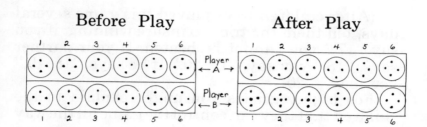

2. Player A takes all his beans from cup 2 and distributes them. Shaded cups show his captures.

Before Play

After Play

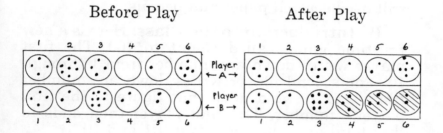

6. COUNT TO 20 (Grades 4-8)

A. Preparation and Materials: None.

B. Introduction to the Class: Try this game with a friend if you finish your work early today. The game is called Count to 20. It has very simple rules. You and your partner take turns counting the numbers from 1 to 20. On your turn you may count either one number or two numbers, starting where the other person left off. The person who counts the number 20 wins. John and Paul, would you come up to the front of the room and try this game?

(After children have played this game several days, tell them the trick to insure winning: if you say "17," you cannot be beaten. If your partner says, "18," you say, "19, 20." If he says, "18, 19," you say, "20."

For higher grades you might explain to the children that they can control the game all the way if they try as early as possible in the game to end their turn on 2, 5, 8, 11, 14, 17, and of course 20.)

7. COUNT TO 100 (Grades 4-8)

A. Preparation and Materials: Children will need scratch paper and pencils.

B. Introduction to the Class: Here is a new free-time game called Count to 100. The first player writes down any number from 1 through 10. The second player writes down any number from 1 to 10, and adds that figure to the first player's number. The first player then adds to that total any number between 1 and 10.

The players in turn continue adding a number from 1 through 10 to the total, and the player who brings the total to an even 100 is the winner.

C. Variation: Begin with 100. Each player in turn **subtracts** a number from 1 through 10. The player who brings the score to an even zero is the winner.

8. TIC-TAC-TOE (Grades 4-8)

A. Preparation and Materials: Children will need pencils and paper.

B. Introduction to the Class: Two of you may enjoy playing Tic-Tac-Toe in your free time. First, draw two horizontal, parallel lines. Then, draw two vertical, parallel lines crossing the first pair of lines.

Example:

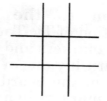

One player will use X's and the other will use O's. On your turn draw your mark in any empty square on the board. The object is to place three of your marks in a vertical, horizontal or diagonal row. If all squares are filled and no one has won, the game is a tie. Play several games, and see who has won the most games when the playing time is over.

Example:

9. MOVING MARKERS (Grades 4-8)

A. Preparation and Materials: Draw a grid, like the one shown below, on paper. Cut 3 paper squares and 3 circles to use as markers.

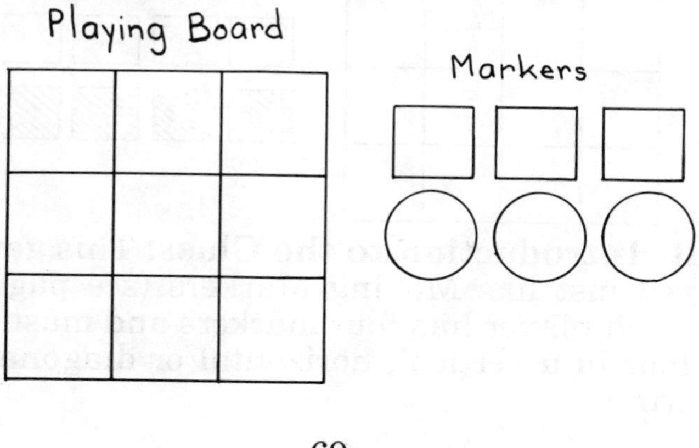

Playing Board

Markers

—69—

B. Introduction to the Class: Moving Markers is a variation of Tic-Tac-Toe. One player will use the square markers and the other will use circles. On your turn lay one of your markers on any empty square of the board. The object is to place three of your markers in a vertical, horizontal or diagonal row.

If all markers have been placed on the board and no one has won, the game continues as follows. Each player in turn moves one of his own markers either up, down or sideways into an adjacent empty square. A marker may not be moved diagonally. This continues until one player gets his three markers in a row in one of the usual Tic-Tac-Toe winning patterns.

10. FOUR IN A ROW (Grades 4-8)

A. Preparation and Materials: Draw a grid, like the one shown in the example, on paper. Cut four red squares and four black squares to use as markers.

Example:

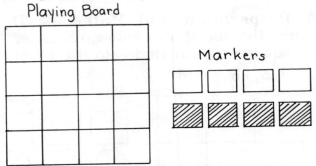

Playing Board

Markers

B. Introduction to the Class: This game is played just like Moving Markers (see page 69), but each player has four markers and must place all four in a vertical, horizontal or diagonal row to win.

11. FIVE IN A ROW (Grades 4-8)

A. Preparation and Materials: Draw a grid, like the one shown in the example, on paper. Cut about 100 red squares and 100 black squares to use as markers. Store the markers in two envelopes clipped to the playing board when the game is not in use.

B. Introduction to the Class: Five In A Row is another version of Tic-Tac-Toe. One player uses red squares for markers and the other player uses black squares. The players take turns placing one marker on an empty square on the board on each turn. No markers are moved once placed on the board.

The object of the game is to place **five** of your markers in a vertical, horizontal or diagonal row.

Example:

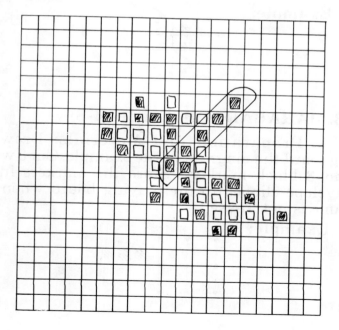

*This activity is available in Challenge Volume I of the **Spice**™ Duplicating Masters.
This activity is available in Challenge Volume II of the **Spice™ Duplicating Masters.

12. FIFTEEN IN A ROW (Grades 4-8)

A. Preparation and Materials: Children will need pencils and paper.

B. Introduction to the Class: This game is played like Tic-Tac-Toe, but you use numbers instead of X's and O's. One player uses only odd numbers from 1 through 9. The second player uses only even numbers from 2 through 10. A player may not use the same number more than once per game.

Each player in turn writes one of his numbers in any square. The object is to complete a diagonal, vertical or horizontal row that adds to 15. Each of the players may have marked some of the numbers in that winning row, but the winner is the person marking the number that brings that row of three numbers to a total of 15.

Example:

13. JARABADACH (Grades 4-8)

A. Preparation and Materials: Draw on tagboard the diagram in the example. Cut 3 white and 3 black markers. Store the markers in an envelope clipped to the playing board when the game is not in use.

Example:

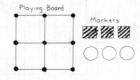

B. Introduction to the Class: Here's an African version of Tic-Tac-Toe. Two players have three markers each. They take turns placing one marker on the board at the intersection of any two or more lines. (Possible placement spots are marked with dots in the example.)

The object is to place three of your markers in a row horizontally, vertically or diagonally. If all markers have been placed on the board and neither player has made a line of three, future plays consist of moving one marker along a line into an adjacent, empty intersection. This continues until one player does move three of his markers into a line.

14. OVID'S GAME (Grades 4-8)

A. Preparation and Materials: Children will need pencils and paper. Each child will need three markers (checkers, squares of paper, etc.).

B. Introduction to the Class: Ovid's Game is another variation of Tic-Tac-Toe. First, draw the playing board.

Example:

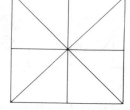

Two players alternate turns. On your turn, place one of your markers on any point where two or more lines meet. You could place your marker in the very center of the board, at any of the four corners, or at the center point of any side of the square.

When all markers have been placed, players take turns moving one of their markers along a connecting line into an adjacent empty point of line intersection. The player who first moves three of his markers into a row vertically, horizontally or diagonally is the winner.

C. Variation: An additional rule may be added if desired; no player may occupy the center of the board until all six markers are placed on the board.

15. LEAP FROG (Grades 4-8)

A. Preparation and Materials: Cut a strip of tagboard and divide it into 9 squares. Cut 4 red markers and 4 black markers from colored construction paper.

Place the red markers on the squares on the left hand side of the board. Place the black markers on the squares on the right side of the board. The center square is empty.

Example:

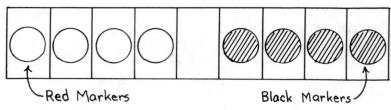

Red Markers   Black Markers

B. Introduction to the Class: This is a free time game for one person. Four red frogs are at one end of the board, and four black frogs are at the other. The object is to move each set of frogs to the opposite end of the board.

Only one frog may be moved at a time. A frog may move into an empty, adjacent space on either side of its present position. Or, a frog may jump any other single frog into an empty space.

How quickly can you leap your frogs to the opposite end of the board? Record the number of moves you took to complete the change. Try the game again, and see if you can do it in fewer moves the next time.

16. 8-MOVE LEAP FROG (Grades 4-8)

A. Preparation and Materials: Divide a strip of tagboard into 5 squares. Place two black markers in the left-hand squares and two red markers in the right-hand squares as shown in the example. The center square is empty.

Example:

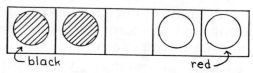

B. Introduction to the Class: Using the same rules for moving as described in Leap Frog (see page 74), try to move each set of markers to the opposite end of the board **in just 8 moves.**

17. COLOR SWITCH (Grades 4-8)

A. Preparation and Materials: Cut a tagboard rectangle. Divide the rectangle into 6 squares. On this playing board place three yellow markers, one red marker and one black marker. Place them in the pattern shown in the example.

Example:

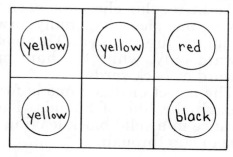

B. Introduction to the Class: The object of this game is to change the positions of the black and red markers on the board.

You may move only one marker at a time, and you move by sliding a marker from one square to any empty square above, below or beside it. You may not move diagonally and you may not jump markers.

Completing this game requires more than 12 moves. Can you do it in less than 20 moves?

18. HARE AND HOUNDS (Grades 4-8)

A. Preparation and Materials: You will need a standard checker board, one black checker and four red checkers. To begin, place the checkers as shown in the example.

Example:

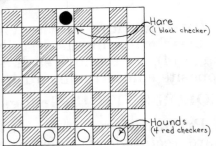

Hare (1 black checker)

Hounds (4 red checkers)

B. Introduction to the Class: One player is the hare. He uses the one black checker. The other player is the hound. He uses the four red checkers.

As in regular checkers, each play consists of moving one piece diagonally forward one square. No jumps are allowed. The hounds may move forward only, while the hare may move either forward or backwards.

The object of this game is for the hare to slip through the line of hounds. The hounds try to advance in a solid block leaving no space for the hare to slip through.

The hare wins if he successfully slips through the line of hounds. He is safe as soon as he passes the last hound, for the hounds cannot move backwards to capture him. If the hounds can corner the hare so he can make no further moves, the hounds win.

19. TRIANGLE RACE (Grades 4-8)

A. Preparation and Materials: You will need a standard checker board, 10 red checkers and 10 black checkers. Arrange the checkers in triangle patterns on the board as shown in the example.

Example:

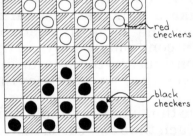

red checkers

black checkers

B. Introduction to the Class: The object of this game is to move your checkers to the opposite end of the board and arrange them in the triangle pattern now occupied by your opponent's checkers. The first player to complete his triangle pattern at the opposite end of the board is the winner.

You move on each turn as in regular checkers, one square diagonally forward. You may jump an opponent's checker to land in an empty square, but the jumped checker is **not** removed from the board. You may move forward only. No checker is ever moved backwards.

Take your time moving pieces across the board. Since a checker may not be moved backward, never move a man farther down the board than needed for the final pattern.

20. CONCENTRATION (Grades 4-8)

A. Preparation and Materials: A standard deck of playing cards.

B. Introduction to the Class: First, lay all the cards face down on the table. Arrange them in several rows. Player one begins by turning any two cards face up. If the two cards are a pair (two fours, two jacks, etc.), he may take those two cards from the table and put them in his pile of "winnings." He may then turn two more cards face up, and continue playing until he turns two cards that do not make a pair. If the two cards do not make a pair, he must turn them face down again on the table in their original positions.

The second player plays in this same way. The game continues until no cards are left on the table. The player who has completed the most pairs is the winner.

While the rules are simple, this is a challenging game. Each time two cards are turned face up, you must try to remember what they were and where they are located on the table. If, for example, you turn a ten face up, you must try to remember if another ten was turned over earlier in the game. If you can remember the location of that other ten, you can easily make a pair.

21. THIRTY-ONE (Grades 4-8)

A. Preparation and Materials: Use all four ace's (an ace counts as 1), 2's, 3's, 4's, 5's, and 6's (24 cards in all) from a standard deck of playing cards.

B. Introduction to the Class: Two players sit side-by-side at a table. Spread all the cards face up on the table.

Players alternately draw one card and add it to a line of cards in front of them. Both players add cards to the same line. A running total is kept of the cards in this line. A player wins if he adds the cards that brings the total to exactly 31, or if his opponent can add no available card without bringing the total to more than 31.

*22. NO CROSSING (Grades 4-8)

A. Preparation and Materials: Children will need pencils and paper. Draw the example on the board.

Example:

B. Introduction to the Class: Two people may play this No Crossing game. On a sheet of paper write the numbers from 1 to 20, scattering them all over the page. Then write the numbers from 1 to 20 a second time, again scattering them all over the page.

The first player draws a line to connect the two number 1's. Then the second player draws a line to connect the two number 2's and so on. But in connecting the number pairs, you may not cross a line already drawn. This, of course, makes connecting number pairs increasingly more difficult as the game progresses. The first player who cannot connect the next pair of numbers without crossing a line loses.

John and Beth, would you please come to the board and use this diagram to play a demonstra-

*This activity is available in Challenge Volume I of the Spice™ Duplicating Masters.

tion game? Their game, completed only as far as matching pairs 1 through 5, might look like this:

*23. CROSS THE BOARD (Grades 4-8)

A. Preparation and Materials: On the board draw a diagram like that shown in the example. Children will need pencils and paper.

Example:

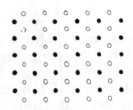

B. Introduction to the Class: Draw a pattern of dots as you see on the board. Two players can use this pattern of dots for a free time game.

One player will use white dots only. The other player uses black dots only. On your turn, draw a line to connect any two of your own dots. Lines may be drawn horizontally or vertically, but **not** diagonally. You may not cross a line already drawn on the playing board.

*This activity is available in Challenge Volume I of the Spice™ Duplicating Masters

The object of the game is to make a continuous path connecting dots of your own color from one side of the board to the other. The white-dot player will be working from top to bottom, or bottom to top of the board. The black-dot player will be working from side to side.

Let's try it once using this diagram on the board. Sue, you may use the white dots. Jack, you will use the black dots. (The finished game might look like this, with Jack, using black dots, winning.)

Example:

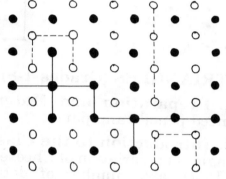

24. CHEESE BOXES (Grades 4-8)

A. Preparation and Materials: Children will need pencils and paper.

B. Introduction to the Class: This is a free time game for two players. Begin by drawing 9 rows of 9 dots each. Each player in turn draws a vertical or horizontal line to connect two adjacent dots. The object is to connect four dots to form a closed square. When your line completes the fourth side of a square, you may put your initials in that square and draw another line.

After all possible lines have been drawn, the winner is the player who has completed the most squares.

*This activity is available in Challenge Volume I of the **Spice**™ Duplicating Masters.
This activity is available in Challenge Volume II of the **Spice™ Duplicating Masters.

Example:

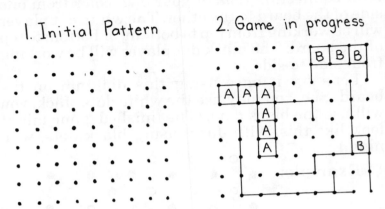

1. Initial Pattern

2. Game in progress

*_{**} 25. TRIANGLES (Grades 4-8)

A. Preparation and Materials: Children will need pencils and paper.

B. Introduction to the Class: The game of Triangles is a variation of Cheese Boxes (see page 81). Place any number of dots in a triangle pattern. Two players alternately draw a line connecting any two dots. Each time a player draws a line to complete the third side of a triangle, he puts his initial inside that triangle and may draw another line. When all dots are connected, the winner is the person who has completed the most triangles.

Example:

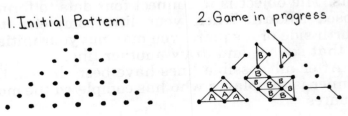

1. Initial Pattern

2. Game in progress

*This activity is available in Challenge Volume I of the Spice™ Duplicating Masters.
**This activity is available in Challenge Volume II of the Spice™ Duplicating Masters

26. SPIDER WEB (Grades 4-8)

A. Preparation and Materials: Children will need pencils and paper.

B. Introduction to the Class: Draw any number of dots on a piece of paper. Two players alternately draw a straight line to connect any two dots. No line may cross a line previously drawn. The winner is the person who draws the last possible connecting line.

Example:

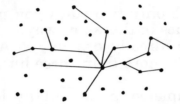

27. DOT EXPLOSION (Grades 4-8)

A. Preparation and Materials: Children will need pencils and paper.

B. Introduction to the Class: Dot Explosion has two players. Begin by drawing three dots in any pattern on your paper. (Demonstrate on the board.)

Example:

The first player draws a line of any shape to connect any two dots and adds a dot anywhere along the line he has drawn. (Demonstrate on the board.)

Example:

←line drawn

←dot added

The second player now connects any two dots and adds a dot on the line he just drew. (Demonstrate on the board.)

Example:

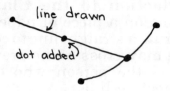

There are only two rules you must follow as play continues in this same way. 1. You may not cross a line already drawn. 2. Any single dot may have no more than three lines connected to it.

The completed game might look like this. (Demonstrate on the board.)

Example:

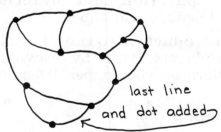

At this point the dot marked with the arrow is the only one not already connected to three lines. There is nowhere else to play. The player drawing the last line is the winner.

C. Variation: Begin with four (or more) dots. The more initial dots, the more difficult the game becomes.

28. CROSS EXPLOSION (Grades 4-8)

A. Preparation and Materials: Children will need pencils and paper.

B. Introduction to the Class: The game of Cross Explosion has two players and is played much like Dot Explosion (see page 83). Begin by drawing three crosses anywhere on your paper. Each player in turn draws a line connecting any two cross arms, and adds a cross bar to the connecting line he just drew.

Example:

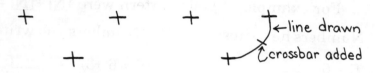

As in the Dot Explosion game, no connecting line may cross a line already drawn. The continued game might look like this.

Example:

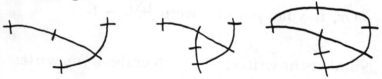

The player making the last possible connection is the winner.

29. GUESS THE PATTERN (Grades 6-8)

A. Preparation and Materials: Use this activity after children have played What Comes Next?, page 125. Children will need pencils and paper.

B. Introduction to the Class: Remember when we played the game, What Comes Next? Numbers were arranged according to a definite pattern and you had to tell what number would come next, if the same pattern were continued. Here is a variation of that game you can play in your free time with a friend.

First, you must think of a definite number pattern. Then, ask your friend to write any five numbers. By each of his numbers write the number that would come next according to your pattern. Use the same pattern for each of his five numbers.

For example, if your pattern were $\boxed{N} + \boxed{N} = 3$

Numbers he writes:		Numbers you write:
1. 6	(6 + 6 +3) =	15
2. 7	(7+7+3) =	17
3. 14	(14+14+3) =	31
4. 15	(15+15+3) =	33
5. 19	(19+19+3) =	41

Or, if your pattern were $\dfrac{\boxed{N}}{2} - 1$

Numbers he writes:		Numbers you write:
1. 2	(2/2–1) =	0
2. 3	(3/2–1) =	½
3. 42	(42/2–1) =	20
4. 16	(16/2–1) =	7
5. 81	(8 1/2–1) =	39½

Your friend will compare the five numbers he wrote with the five numbers you wrote and try to guess your pattern.

30. CLEAR THE BOARD (Grades 6-8)

A. Preparation and Materials: Cut a triangle of tagboard. On the triangle mark a pattern of 15 circles as shown in the example. Cut 14 paper markers.

Example:

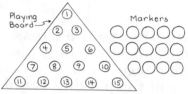

B. Introduction to the Class: To begin this game, put a marker on top of every circle except the top circle (number 1) in the pyramid.

To play, jump any one marker over any other adjacent marker into an empty space. Remove the jumped marker from the board. For example, you could begin by jumping marker number 4 over marker number 2 and into the empty space number 1. Remove the jumped marker number 2 from the board.

The object of this game is to continue jumping and removing markers from the board until only 1 marker is left.

**31. JUMP (Grades 6-8)

A. Preparation and Materials: Draw a grid, like the one shown in the example, on tagboard. Color the squares shaded in the example. This aids children in placing the markers to set up the game.

Provide 32 markers (beans, checkers, paper squares, etc.). Put the markers in an envelope clipped to the playing board when the game is not in use.

Example:

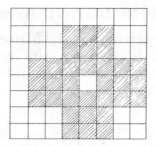

B. Introduction to the Class: Here is a new free time game one person may play alone. To set up the playing board place a marker on each colored square of the playing board.

Jump one marker over any adjacent marker either horizontally or vertically, landing in an empty space. Remove the jumped marker from the board.

Continue in this way until only one marker remains on the board. This last marker must end in the center square.

32. REVERSE (Grades 6-8)

A. Preparation and Materials: Use a standard checker board or draw a grid, like the one shown in the example, on heavy paper. Cut 64 tagboard circles for markers. Color one side of each circle black. Leave the other side its natural color. Put the markers in an envelope clipped to the board when the game is not in use.

B. Introduction to the Class: To begin the game of Reverse, each of two players takes 32 markers. One player will use his markers with the black side up, and the other will use his with the white side up.

On the opening four plays, the players alternately place one marker of their color on one of the four center squares of the board.

Example: Opening Four Plays

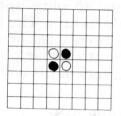

On subsequent plays a player must place his marker adjacent to a marker of the opposite color. It can be in a vertical, horizontal or diagonal line with the marker of the opposite color. Also, the marker must be placed so as to "trap" the opponent's marker between two or more markers of the player's own color. The trapped marker is now turned over to become the opposite color.

Example:

1. A white marker is trapped between two black markers.

2. The trapped white marker is turned over and becomes black.

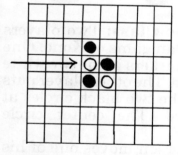

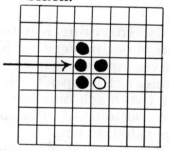

If on your turn there is no place to play where you can trap an opponent's marker, you lose your turn. Your opponent continues to play until there is an opportunity for you to re-enter the game by trapping an opponent's marker.

When one player has used all his markers, the other may continue to play until no remaining plays are available to him. The game ends when all markers have been used or when no further plays are available to either player. The person with the most markers of his color on the board is the winner.

33. KONO (Grades 6-8)

A. Preparation and Materials: Draw a diagram like that shown in the example on tagboard. Cut two black and two white markers. Store the markers in an envelope clipped to the playing board when the game is not in use.

Example:

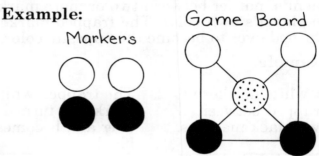

Markers

Game Board

B. Introduction to the Class: Two players can play this very old Korean game of Kono. One player puts his two white markers in the two white circles at the top of the board. The other player puts his two black markers in the two black circles at the bottom of the board. The center circle (speckled) is empty.

On the first play one person moves one of his markers into the center circle. The second player then moves one of his markers into the circle just made empty. The players alternate turns, and on each turn move one marker along a line into the empty circle.

34. 4-WAY JUMP (Grades 6-8)

A. Preparation and Materials: Draw on tagboard the diagram shown in the example. Cut eight black and eight white markers. Store the markers in an envelope clipped to the game board when the game is not in use.

Example:

Playing Board   Markers

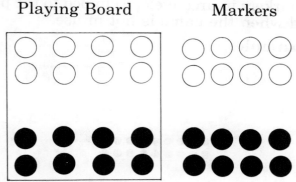

B. Introduction to the Class: Two players can play this Korean version of checkers. One player puts his white markers on the white circles on the board, and the other player puts his black markers on the black circles.

Players alternate turns. A move consists of moving one marker either vertically or horizontally (**not** diagonally) from its present position into an empty, adjacent space. Or you may jump one of your own markers to land in a space occupied by your opponent. In this case the opponent's marker is "captured," and is removed from the board. You do not have to jump when a jump is available, and you cannot jump unless a capture is made.

If a marker is surrounded by opponent's markers so it cannot move in any direction, it also is captured and removed from the board.

The winner is the person with markers still on the board after all the opponent's markers have been captured.

35. SCISSORS CHESS (Grades 7-8)

A. Preparation and Materials: Draw a grid, like the one shown in the example, on tagboard. Cut 18 black markers and 18 white markers. Store the markers in an envelope clipped to the playing board when the game is not in use.

Example:

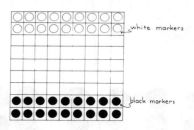

B. Introduction to the Class: The object of this game is to be the first player to arrange any five of your markers in a vertical, horizontal or diagonal line. Original starting positions do not count in the line of five.

To begin the game, place all your markers in the first two rows of your side of the board (see example). Players alternate turns. On your turn you may move any marker any number of free spaces vertically or horizontally. You may not land in an occupied square. To "jump" an opponent you must on one turn move into an adjacent square with that marker. On your next turn you may jump over him vertically or horizontally into an empty space. The jumped man is removed from the board.

You may also remove an opponent's marker by catching him "in the scissors." This means you have moved two of your markers on opposite sides of his marker, and in squares adjacent to that marker, either vertically or horizontally. If you voluntarily move one of your markers between two of your opponent's markers, you are **not** caught "in the scissors."

And don't forget the object of this game. While you are busy making captures and avoiding being captured, you are also trying to line up five of your markers in a vertical, horizontal or diagonal line.

The winner is the person who first lines up five of his markers. A player forfeits the game when he has less than five markers remaining on the board, for he cannot possibly form a line of five.

36. THE MILL (Grades 7-8)

A. Preparation and Materials: You will need 9 red markers and 9 black markers. The markers can be checkers, squares of paper, or any other similar objects.

To make the playing board draw three concentric squares on tagboard. The smallest square should be about 6 inches long on each side. Draw the lines connecting the center square to the outside square.

Example:

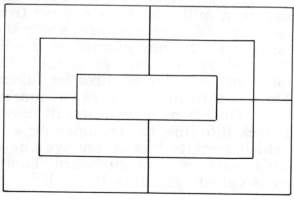

B. Introduction to the Class: Two players have 9 markers each. The players take turns laying one marker on the board on any of the corners. A corner is any intersection of vertical and horizontal lines.

Example: Corners are marked with an X.

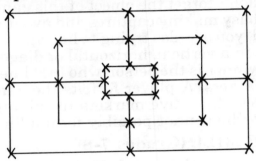

After all markers are placed on the board, play continues by each player in turn moving one marker from its present position along a connecting line into an adjacent, empty corner.

The object of the game is to create a "mill" and remove your opponent's markers from the board. A mill is a line of three markers of your own color placed in a horizontal, vertical or diagonal row of adjacent corners.

Whenever you successfully align three of your men in a mill, you may remove from the board any one of your opponent's markers except one of a set he has aligned in a mill.

After a player establishes a mill, he can continue using this line of three for future captures. On one turn he can move one marker out of the line. On his next turn he can move that marker back into line, thus completing a line of three, which permits him to remove one of his opponent's markers from the board. This series of plays is called "grinding the mill."

Grinding the mill is a desirable play, but it has its hazards. When you move a marker out of line, each of those three markers is now vulnerable to capture by your opponent if he creates a mill on his next play.

You may establish more than one mill at a time. Then, if one mill is destroyed by a capture, you will have another already aligned and ready to grind.

When a player is left with fewer than three markers on the board, it is no longer possible for him to create a mill and make further captures. His opponent has then won the game.

CHAPTER III:
"Puzzles and Brain Teasers"

Problem solving gains challenging appeal when presented in the form of puzzles and brain teasers.

1. SWITCHEROO (Grades 4-8)

A. Preparation and Materials: Each child will need a scrap of paper and a pencil.

B. Introduction to the Class: Tear three small scraps of paper. Put the three scraps in a row on your desk. Put an X on the middle scrap. (Demonstrate.)

Example:

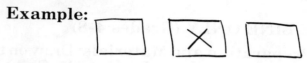

The object of this puzzle is to place one of the outside papers into the center position without moving the paper already in that center position.

Solution: Move paper #1 to the far right. Now, paper #3 becomes the center, and the original center piece, #2 has not been moved.

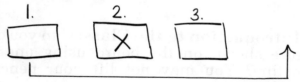

2. ALTERNATE GLASSES (Grades 4-8)

A. Preparation and Materials: Arrange three full glasses of water and three empty glasses in a row as shown in the example.

Example:

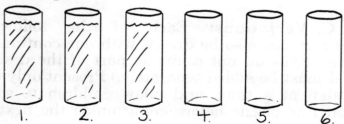

B. Introduction to the Class: Look at this row of glasses. The first three glasses in the row are full of water. The last three are empty. Can you move **one glass only** so the row will alternate empty and full glasses?

Solution: Pick up glass 2. Pour its contents into glass 5 then place glass 2 back in its original position.

**3. ONE LINE ONLY (Grades 4-8)

A. Preparation and Materials: Draw on the board the figure shown in the example. Children will need scratch paper and pencils.

Example:

B. Introduction to the Class: Can you draw the figure shown on the board using one continuous line? You may not lift your pencil or retrace a line.

Solution:

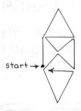

start →

C. Variations: Each of the following designs can also be drawn with one continuous line. (Eyes do not count as part of the design, and must be added separately.) Present only one design at a time, and allow children to solve that line puzzle before going on to the next.

This activity is available in Challenge Volume II of the **Spice™ Duplicating Masters.

Problem   Solution

1.

1.
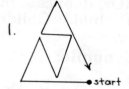
● start

2.

2.

● start

3.
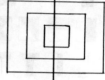

3.
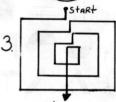
● start

4.

4.
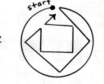
start

5.

5.
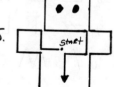
start

6.
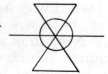

6.
start →

4. RELATIONSHIPS (Grades 4-8)

A. Preparation and Materials: Place on the board figures like those shown in the example. (Or, duplicate material making one copy for each child.) Children will need pencils and paper.

Example:

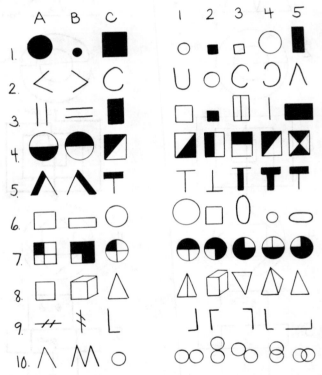

B. Introduction to the Class: Look at the first two figures on the board. In what way are the first two figures alike, Joe? Yes, both are black circles, but B is smaller than A. Now look at the next figure. It is a large black square. Which of the remaining figures in that row is related to the

—102—

square in the same way the second circle was related to the first? Yes, Tammy, the small black square. The large black circle is related to the small black circle in the same way that the large black square is related to the small black square. On your paper by the number 1, write "2," to show the second figure is the correct answer.

Now, look at the next two figures. Think how the first two figures are related to each other. Now, look at the third figure. In that same row can you find a figure related to this in the same way the first two are related? Yes, Bill, figure "4." Why is this the correct figure? (Figures A and B are alike, except reversed. Figures C and 4 are also alike, except reversed.)

Finish the work in this same way. Decide how the first two figures in each row are related to each other. Match the third figure with another figure in that row that is related in the same way the first two figures are related.

5. CHANGE FOR A DOLLAR (Grades 4-8)

A. Preparation and Materials: None.

B. Introduction to the Class: How could you make change for a dollar bill using exactly 50 coins?

Solution:
1. 45 pennies, 1 quarter, 2 dimes and 2 nickels.
2. 40 pennies, 2 dimes and 8 nickels.

6. SHORTEST ROUTE (Grades 4-8)

A. Preparation and Materials: Draw on the board the diagram shown in the example.

Example: A   B

C

B. Introduction to the Class: The cities of A, B and C decided they needed a highway to allow people from any one city to travel directly to any other of the three cities. Because the highway budget was limited, they had to build the shortest possible highway to connect the cities. What route did they select for this highway?

Solution:

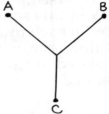

7. GEOMETRIC JIGSAW PUZZLES (Grades 4-8)

A. Preparation and Materials: Cut a basic geometric shape (circle, square, oval, rectangle, etc.,) from construction paper or tagboard. Lightly pencil in several straight lines which will serve as cutting guides. Copy the shape and cutting lines onto a sheet of paper which will serve as an answer guide.

Cut the shape along the lines drawn. Put the pieces in an envelope and on the envelope draw the shape which can be made from these pieces.

Repeat this process until you have made a group of puzzles. Put all the envelopes in a box. Tape the answer guide to the bottom of the box.

Example:

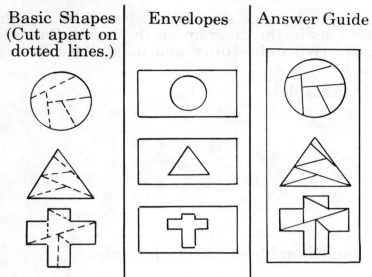

| Basic Shapes (Cut apart on dotted lines.) | Envelopes | Answer Guide |

B. Introduction to the Class: In this box are several envelopes. On each envelope is drawn a shape. In the envelope are a number of pieces that can be fitted together like a jigsaw puzzle to form the shape you see on the envelope. Take out the pieces from one envelope at a time. See if you can arrange the pieces to make the shape shown on the envelope. If you are really stumped, the answers are shown on a sheet taped to the bottom of the box. But don't peek until you have **really** tried to solve the puzzle.

Be sure to put all the pieces back into the envelope before opening another puzzle.

8. COIN TRICKS (Grades 4-8)

A. Preparation and Materials: Draw on the board the diagrams shown in the example. Provide circles cut from heavy paper for children to use for coins.

B. Introduction to the Class:

1. Arrange six coins to form two straight lines, as in the diagram on the board. Can you move **two coins only** and make them form a circle?

Example:

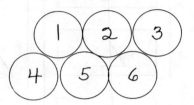

Solution: Move 5 to touch 4 only.

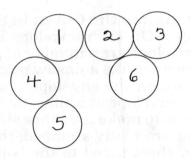

Move 3 to touch 5 and 6.

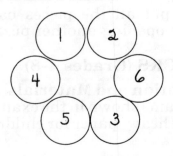

2. Use the same initial pattern as shown in the example. Can you form a circle by moving **three** coins?

Solution: Move 6 to touch 4 and 5.

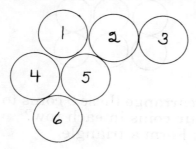

Move 5 to touch 3 and 2.

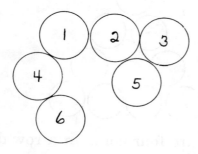

Move 3 to touch 5 and 6.

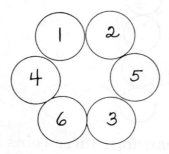

3. Here are three rows with three coins in each row.

Can you rearrange these 9 coins to make three rows with **four** coins in each row?
Solution: Form a triangle.

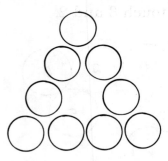

4. Here are four coins in a row down, and three coins in the row across.

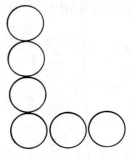

Can you rearrange these coins to make four coins in each line?

Solution: Take the upper coin from the vertical row. Place it on top of corner coin. Now, each row totals 4.

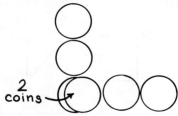

9. TOOTHPICK TRICKS (Grades 4-8)

A. Preparation and Materials: Put on the board the problems given from 1 to 6. You may want to use only one each day or present them all at once. Provide a box of toothpicks for children to use while working on the solutions to these problems.

B. Introduction to the Class: Here are some toothpick puzzles. The directions are given. Think over these problems in your free time today. You may take toothpicks from this box to test your ideas for solutions to these problems. We will discuss them together just before dismissal time today. Do you think you will have a solution?

1. Problem: Take away two toothpicks to make this statement true:

Solution:

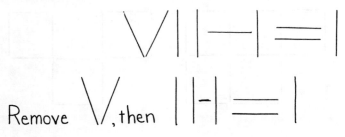

2. Problem: You are given 9 toothpicks. Can you make ten of them? No fair breaking them!
Solution:

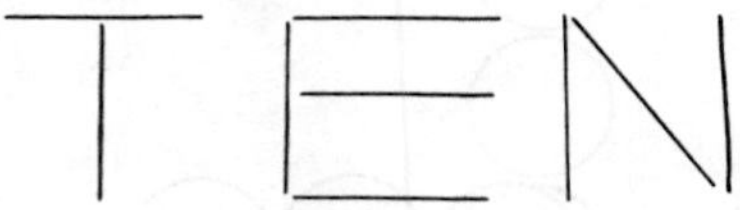

TEN

3. Problem: Move (not remove) two toothpicks to make this statement true.

Solution:

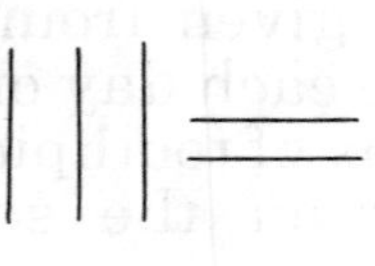

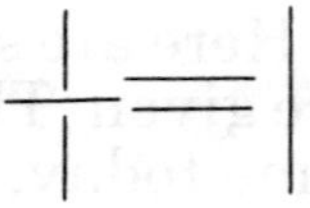

$$||| = |$$

$$+ = | \quad \text{or} \quad \lfloor = |$$

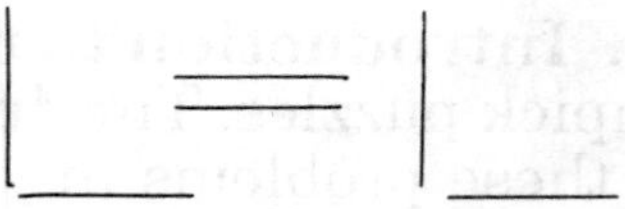

4. Problem: Here are 15 toothpicks arranged to form 5 squares. Can you remove three toothpicks and have 3 squares left?

Solution:

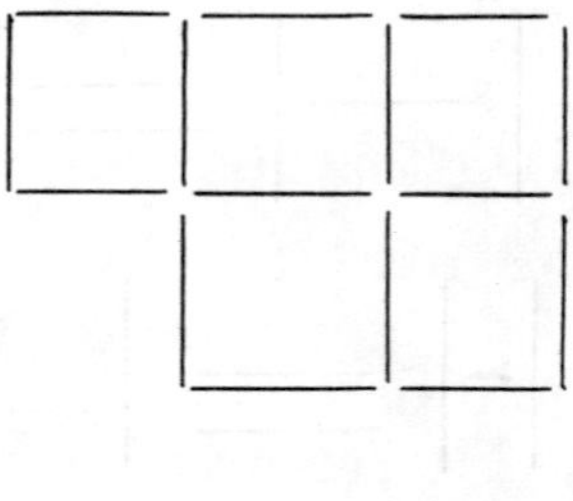

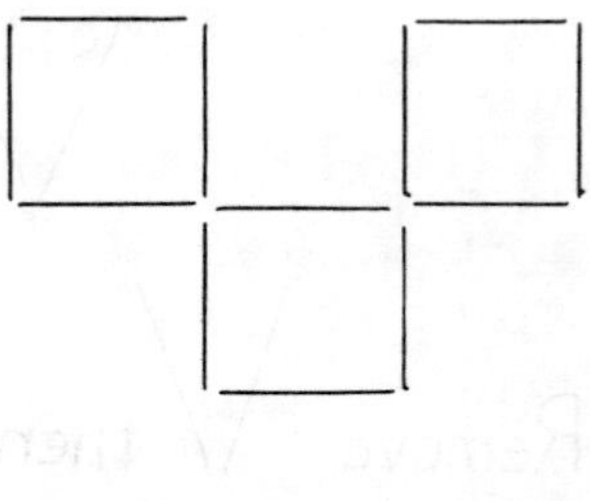

5. Problem: Here are 9 toothpicks which form 3 triangles. Can you move just 3 toothpicks and form 5 triangles?

Solution:

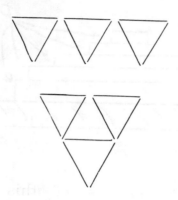

(The outer rim of the figure shown in the solution counts as the 5th triangle.)

6. Problem: You have 8 toothpicks. Can you take away 3 and still have 8 left?
Solution:

10. CHINESE PYRAMID (Grades 4-8)

A. Preparation and Materials: You will need 3 blocks of wood about 4 inches square. Cut 5 discs of tagboard. One should be about 3½ inches in diameter, and the others progressively smaller so the discs can be stacked to form a tower.

Example:

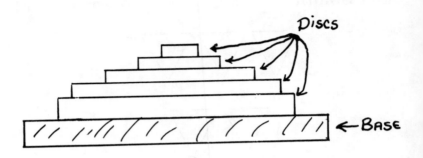

Commercial versions of this game are available. They consist of a triangular base with three upright dowels. The discs are stacked on the dowels. Most come with 8 discs. If you use this, remove 3 discs. Moving 8 discs according to the rules of this game requires a minimum of 255 moves, which is beyond the patience level of most children.

B. Introduction to the Class: This is a puzzle that has fascinated people for centuries. To play, you place the three blocks in a row on your desk. On one block, stack all 5 discs, with the largest on the bottom and the smallest on top.

The problem is to move all the discs to make the pyramid on either of the other blocks. You may move only one disc at a time, and you may place it on either of the other two blocks. You may never place a larger disc on top of a smaller disc.

This puzzle requires much patience. The five discs cannot be transferred to another block in less than 31 moves!

11. WHICH WEIGH? (Grades 4-8)

A. Preparation and Materials: Make sure children understand how a balance scale works. Use the illustration to aid your discussion.

Example:

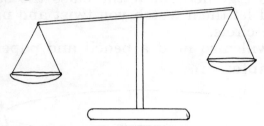

B. Introduction to the Class: Listen carefully to this problem. Think about a solution whenever you have free time today. At dismissal time I will see if any of you have discovered the solution.

Problem: A man gave his son a balance scale and 12 coins which all looked and felt alike. He said, "One of these coins is heavier than the others. If you can find the heavy coin in just three weighings, you may keep the coins."

If you were the son, could you have kept the coins?

Solution:

1. Divide the coins into 3 groups of 4 coins each. Put one group in each of the two cups of the scale.

2. Divide the heavy group in half, putting 2 coins in each scale cup. (If the first two groups balanced, the heavy group was the one not yet weighed.)

3. Divide the heavy group in half, putting one coin in each scale cup to discover which is the heavy coin.

12. MOEBIUS STRIP (Grades 4-8)

A. Preparation and Materials: Cut a strip of paper about 18 inches long. A length of adding machine tape is perfect. Make a single twist in the strip and then paste or staple the two ends together. Show the class the straight strip and let them watch you twist and paste to form the circle.

You will also need a pencil and paper.

Example:

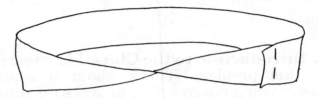

B. Introduction to the Class: I have a magic circle of paper. This paper has only one side; I will prove it to you.

If you drew a line on ordinary paper, beginning on one side and continuing to the other side, you would have to cross an edge, wouldn't you? But on this paper I can draw a line going from one side to the other without crossing an edge. So therefore it must have only one side! Watch while I show you. (Put your pencil on the center of the strip. Draw a continuous line down the center of the strip. You will eventually come back to the place where you started. The line will be drawn on both sides of the paper, but you will not have crossed an edge.)

My circle is magic in other ways, too. What do you think will happen if I cut along the line I just drew? (Let the children guess.) Let's find out. (Cut along the line. Instead of getting two separate loops, as they might suspect, you will get one large loop with a double twist.)

What do you think will happen if I cut this big loop in two again? (Let them guess.) Let's find out. (Cut the loop in two, lengthwise. Instead of getting one even larger loop, as the children might have guessed because of their last experience, you will get two linked loops.)

**13. MAGIC FIGURES (Grades 4-8)

A. Preparation and Materials: Place on the board one or more of the problems from 1 to 3. Children will need scratch paper and pencils for figuring.

B. Introduction to the Class: In mathematics class you have studied magic squares. You know that a magic square is one in which all rows of numbers across, down, or vertically add to the same total. On the board are some problems involving magic figures. See if you can solve these problems in your free time today.

1. Problem: This figure is magic in three ways. Can you find the ways?

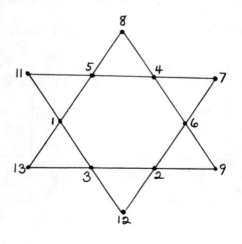

This activity is available in Challenge Volume II of the **Spice™ Duplicating Masters.

Solution: 1. The sum of each small triangle is 17. 2. The sum of the points on each large triangle, is 30. 3. The sum of each line of four numbers is 27.

2. Problem: Place the numbers 1 through 9 at a dot along the rim of this triangle so that the sum of each side of the triangle is equal.

Problem: Solution:

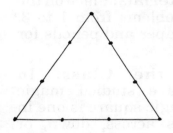
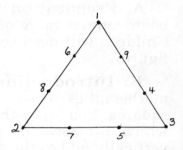

3. Problem: Fill in the spaces in this figure with the numbers 1 through 9, using each number only once, to make a magic square which totals 15 when added in any direction.

Problem: Solution:

4	9	2
3	5	7
8	1	6

14. NUMBER MEMORY (Grades 4-8)

A. Preparation and Materials: Using a felt pen or black crayon, draw on a sheet of unlined paper one of the number groups in the example. Put each diagram on a separate sheet. Make sure each drawing is large enough to be seen by all the children when you hold it in front of the class.

The examples shown range in order from simpler to more complex. The short explanation of the pattern under each diagram is for your convenience in leading a class discussion after all the patterns have been shown. Children will need paper and pencils.

Example:

3	6	9
4	7	10
5	8	11

Row 1 down increases by 1. Each row across increases by 3.

2	4	6
4	6	8
6	8	10
8	10	12

Row 1 down increases by 2. Each row across increases by 2.

10	8	6
9	7	5
8	6	4
7	5	3

Row 1 down decreases by 1. Each row across decreases by 2.

3	5	7
6	8	10
9	11	13
12	14	16

Row 1 down increases by 3. Each row across increases by 2.

2	8	4
4	16	8
6	24	12
8	32	16

Row 1 down increases by 2. Row 2 down is 4 times row 1. Row 3 down is ½ row 2.

3	12	6	7
4	16	8	9
5	20	10	11
6	24	12	13
7	28	14	15

Row 1 down increases by 1. Row 2 down is 4 times row 1. Row 3 down is ½ row 2. Row 4 increases by 2.

B. Introduction to the Class: I will hold up a pattern of numbers for you to study carefully for one minute. The numbers are arranged according to a definite pattern. See if you can decide what that pattern is.

At the end of one minute I will put down my sheet and say, "Go." Then you may take your pencils and try to draw the figure you saw, putting each number exactly where you saw it on my sheet. If you have figured out the pattern of numbers, it will be very easy for you to make an exact copy.

After two minutes, I will say, "Stop." Put your pencils down immediately. Let's see how many can draw an exact copy from memory.

(After the class has completed each drawing, and work has been checked, show the original diagram again. Use the notes below the illustrated examples to aid you in leading a discussion concerning the pattern which the numbers follow.)

15. PASSING TRAINS (Grades 4-8)

A. Preparation and Materials: Draw on the board the diagram shown in the example. Leave this diagram on the board during the day for children's reference.

Example:

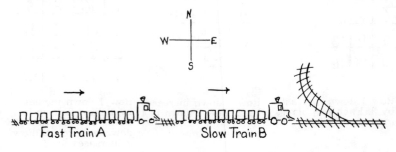

B. Introduction to the Class: Here is a brain teaser to think about in your free time today. At dismissal time I will see if any of you have figured it out.

Here is the problem: A fast express train overtook a slow freight train on a single track near a short siding. The fast train wanted to pass the slow train.

The railroad siding was short. It would hold only 5 cars (or 4 cars and 1 engine) at a time. The fast train had 12 cars. The slow train had 9 cars. What is the quickest way for the fast train to pass the slow train?

Solution: Slow train B moves past the siding, backs into the siding, uncouples its 5 end cars, and then moves back out along the main track towards the east.

Fast train A moves past the siding, backs in, picks up the 5 cars from Train B, then moves out onto the main track and backs to the west of the siding.

Slow train B backs into the siding.

Fast train A passes the siding, leaving the 5 end cars from Train B just west of the siding, and continues on its way.

Slow train B then merely has to move onto the main track, back into its 5 end cars, recouple them, and continue on its way.

16. PROBLEMS IN ARRANGEMENT AND CONSTRUCTION (Grades 4-8)

A. Preparation and Materials: Place on the board one or more of the problems from 1 to 3.

B. Introduction to the Class: Here are some problems you may work on in your free time today. You may want to cut small pieces of paper, number them as shown on the board, and use them in trial arrangements as you attempt to solve these problems. At dismissal time I will ask for your solutions.

1. Problem: Arrange these cards in pairs so that all pairs will have the same total.

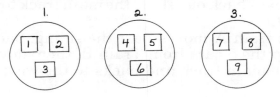

Solution: All pairs total 26. Thus the pairs should be 2-24, 4-22, 6-20, 8-18, 10-16, 12-14.

2. Problem: Remove a block from one circle, and place it in a new circle so that the total of the numbers in each circle will be equal.

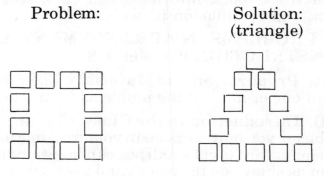

Solution: Move the number "9" from circle 3 to circle 1. Now all circles total 15.

3. Problem: Rearrange the blocks in this square to form a figure with five blocks to a side.

Problem: Solution:
                            (triangle)

*⁎⁎17. FIND THE NUMBERS (Grades 4-8)

A. Preparation and Materials: Draw on the board a circle, square, triangle and rectangle. Allow each shape to partly overlap the others. Number each section of the diagram formed by crossing lines.

Under the diagram, write the questions pertaining to the numbered sections. Children will need scratch paper and pencils.

Example:

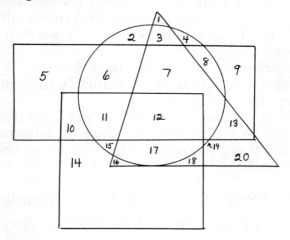

What numbers are:

1. In the rectangle, but not in the circle, square or triangle? (5, 9)

2. In the triangle, but not in the rectangle or square? (20, 19, 3, 1)

3. In the square, but not in the circle or triangle? (10, 14)

4. In the circle, but not in the triangle or rectangle? (15, 2, 4)

—121—

*This activity is available in Challenge Volume I of the **Spice**™ Duplicating Masters.
This activity is available in Challenge Volume II of the **Spice™ Duplicating Masters.

5. In the rectangle, but not in the triangle or square? (5, 6, 8, 9)

6. In the square, but not in the rectangle or circle? (14, 16, 18)

7. In the triangle, but not in the circle or square? (20, 13, 1)

B. Introduction to the Class: I want to see how thoroughly you can pay attention to details and follow directions. Look at the figures on the board. What four basic shapes do you see, Tim? Yes, a circle, square, triangle and rectangle. Each figure overlaps the others. Each section has a number.

Number your papers from 1 to 7. By each number, write the section numbers which correctly answer the questions asked.

You may have 5 minutes to work. At the end of that time we will check your work and discuss each part together. Let's see who can have the most correct answers.

18. GUESSING A BIRTHDAY (Grades 4-8)

A. Preparation and Materials: None.

B. Introduction to the Class: Here is a trick you can play on a friend. Ask him to write down the month and day of his birthday, using 1 for January, 2 for February, etc. Don't let him show you what he wrote.

Ask him to multiply the month of his birth by 5, add 6 to that answer, multiply by 4, add 9, and multiply by 5. Then ask him to add the day of his birth to that total. Have him tell you just the total, and you can tell him the month and day of his birth.

All you have to do is subtract 165 from the

total he gives you. The last two numbers of the answer will be the date of his birth. The first number (or numbers) will tell you the month of his birth.

Practice this with your classmates. When you have mastered the process. Try it on your family and friends at home.

Example: Birthdate Sept. 22

$$
\begin{array}{r}
9 \\
\times\ 5 \\
\hline
45 \\
+\ 6 \\
\hline
51 \\
\times\ 4 \\
\hline
204 \\
+\ 9 \\
\hline
213 \\
\times\ 5 \\
\hline
1065 \\
+\ 22 \\
\hline
1087 \\
-\ 165 \\
\hline
922
\end{array}
$$

9th month (Sept.) 22nd day

19. MAKE A SQUARE (Grades 5-8)

A. Preparation and Materials: Cut 6 squares of paper. Use these to make **six different** puzzles, according to these general directions.

1. Cut a square into two irregular pieces.

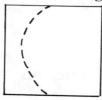

2. Lay the pieces together, with two **straight** edges touching.

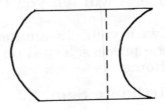

3. Trace this new shape onto tagboard and cut it out.

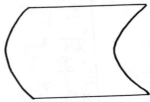

The shapes you cut might look like those in the example. Dotted lines indicate where these shapes could be cut to form the original squares.

Example:

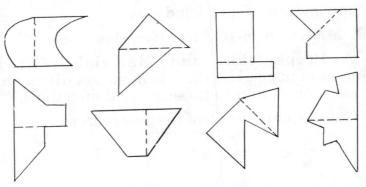

Put these shapes into an envelope labeled "Make a Square."

B. Introduction to the Class: Here is a different kind of puzzle you may work on in your free time. Each of the shapes in this envelope could form a square if you made one straight cut with scissors, and then re-arranged the two pieces. You may take this envelope to your desk and see if you can figure out where that cut should be made, and how the two pieces could be re-arranged to form a square.

Try to think it through in your head. If you are really stumped, trace the shape on scratch paper and try cutting your scratch paper pattern.

At the end of the week we will discuss this puzzle together. I will call on people to cut each shape and re-arrange the two pieces to form a square.

* ** 20. WHAT COMES NEXT (Grades 5-8)

A. Preparation and Materials: Write on the board a series of numbers that follow a definite pattern in their relationship to each other. After each series, write a choice of answers for the number which would come next if the pattern were continued. In the example, the pattern of each series is given, and the correct answer is circled. This will aid you in leading a class discussion following the work period. Children will need scratch paper and pencils.

Example:

1. 7   8   10   11   13   14   14   15 (16) 18   19
      +1 +2 +1 +2 +1   (+2)

2. 9   11   14   18   23   29   30   34 (36) 39   40
      +2 +3 +4 +5 +6   (+7)

*This activity is available in Challenge Volume I of the Spice™ Duplicating Masters.
**This activity is available in Challenge Volume II of the Spice™ Duplicating Masters.

3.	2	4	2	6	2	8		②	4	6	8	10
----	---	---	---	---	---	---	---	-6-	---	---	---	---
		+2	−2	+4	−4	+6		−6				

3. 2   4   2   6   2   8   (2) 4   6   8   10
        +2 −2 +4 −4 +6   −6

4. 3   5   7   9   7   5   (3) 5   7   9   11
        +2 +2 +2 −2 −2   −2

5. 14 15 13 16 12 17   9 (11) 15 16 17
        +1 −2 +3 −4 +5   −6

6. 90 45 50 25 30 15   5 10 15 (20) 35
        ÷2 +5 ÷2 +5 ÷2   +5

7. 40 20 24 12 16   8   4   8 (12) 16 20
        ÷2 +4 ÷2 +4 ÷2   +4

8. 5   6   9 15 16 19   20 21 22 23 (25)
        +1 +3 +6 +1 +3   +6

B. Introduction to the Class: Here is an interesting exercise in logical thinking. On the board you see groups of numbers in a series. These numbers are arranged according to a definite pattern. If you continued the pattern in series 1, which of the numbers in the second group would come next? To find the number, you must first decide what the pattern is. I will help you with the first series.

The first number is 7 (point to 7). Add 1 to get 8 (point to 8). Add 2 to get 10 (point to 10). Add 1 to get 11 (point to 11). Add 2 to get 13 (point to 13). Add 1 to get 14 (point to 14). So the pattern is: add 1, add 2, add 1, add 2, and so on. What must you do to 14 to get the next number in the series? Yes, Dick, add 2. That would make 16 (circle 16 in the answer column).

Number your paper from 1 to 8. Beside each number write the number which would come next if that series were continued. I will give you four minutes to work. At the end of that time we will check your answers and discuss the patterns you discovered in each series.

**21. PUZZLE IN THE ROUND (Grades 6-8)

A. Preparation and Materials: Draw on the board 10 circles around a center circle. Draw connecting spokes as shown in the example.

Example:

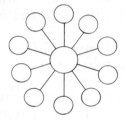

B. Introduction to the Class: Try this mathematics puzzle in your free time today. Copy the pattern you see on the board. Write one of the numbers from 1 through 11 in each circle to make each row of 3 circles add to the same sum.

Solution: (Each row of 3 circles adds to the sum of 18.)

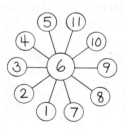

This activity is available in Challenge Volume II of the **Spice™ Duplicating Masters.

22. MATCH GAME (Grades 6-8)

A. Preparation and Materials: None.

B. Introduction to the Class: One day Jane White, Amy Black and Beth Brown were walking to school. "Look at our skirts," said one girl. "One is white, one is black and one is brown, just like our names."

"Yes," said the girl with the black skirt, "but none of us is wearing a skirt color to match our own name."

"I noticed that too," said Jane White.

What color skirt is each girl wearing?

Solution:

    Jane White: brown skirt

    Amy Black: white skirt

    Beth Brown: black skirt

Jane White isn't wearing white, for that would match her name. She isn't wearing black, because she spoke with the girl wearing black. So her skirt has to be brown.

Amy Black can't be wearing black to match her name. She can't be wearing brown because Jane White's skirt is brown. So she must be wearing white.

That leaves the black skirt for Beth Brown.

**23. FOXY FENCING (Grades 6-8)

A. Preparation and Materials: Draw on the board the diagram shown in the example.

Example:

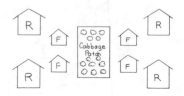

This activity is available in Challenge Volume II of the **Spice™ Duplicating Masters.

B. Introduction to the Class: Four rabbit families built their homes around a large cabbage patch. The rabbit houses are labeled "R" in the diagram on the board. For weeks they ate cabbages and enjoyed a good, safe life.

But one dark night four fox families moved into the neighborhood building their homes close to the cabbage patch. The foxes' houses are labeled "F."

Now, it was no longer safe for the rabbits to go to the cabbage patch, so they decided to build a strong, high fence. They wanted to build it so their four homes, plus the cabbage patch, would be **inside** the fence, and the four foxes homes would be **outside** the fence. How can they build the fence?

Solution:

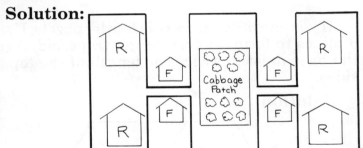

24. TRAPEZOID PUZZLES (Grades 6-8)

A. Preparation and Materials: Cut 9 trapezoids from paper following the dimensions in the example. Put the 9 trapezoids in an envelope labeled "Trapezoid Puzzle."

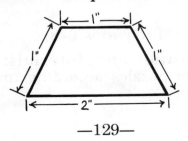

B. Introduction to the Class: Here's a puzzle to try in your free time. This envelope contains 9 trapezoid shapes. (Show one shape to the class.) The bottom of each trapezoid is twice as long as the top or sides.

First, take just 4 trapezoids from the envelope. Can you arrange them to form a larger trapezoid with the base twice the length of the top and sides?

Solution:

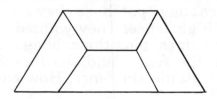

If you can do that, try this harder puzzle. Use all 9 pieces to form an even larger trapezoid. Again, the base must be twice the length of the top and sides.

Solution:

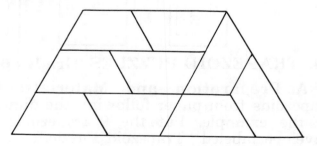

**25. TANGRAM (Grades 6-8)

A. Preparation and Materials: Make one or more tangram puzzles according to these directions:

This activity is available in Challenge Volume II of the **Spice™ Duplicating Masters.

1. Use a ruler to draw a square on heavy paper. Draw a line to connect corners D and B.

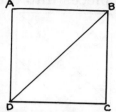

2. Mark the mid-points between D and C (labeled E), and C and B (labeled F). Draw a line to connect E and F.

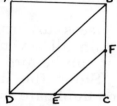

3. Mark the mid-points between D and B (labeled G), and E and F (labeled H). Draw a line to connect corner A with points G and H.

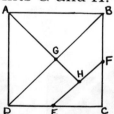

4. Mark the mid-points between D and G (labeled I), and G and B (labeled J). Draw a line to connect points I and E. Draw a line to connect points J and H.

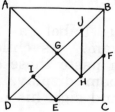

5. Cut along all lines to make a 7 piece tangram puzzle. Put the 7 pieces into an envelope labeled "tangram."

B. Introduction to the Class: Here is a new puzzle you may use in your free time. This puzzle is called a tangram and is made of a square cut into seven pieces.

Arrange the pieces to form pictures. Each picture must contain all seven pieces. Here are some pictures that can be formed with the seven tangram pieces. (Show the illustration in the example.) Can you duplicate these pictures? Can you create new pictures of your own?

Example:

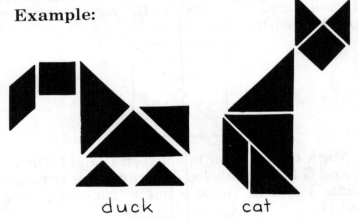

duck   cat

C. Variation: Label a large sheet of tagboard "Tangram." As children create new designs, let them paste these designs onto the chart for classroom display. Each child may title his design and sign his name.

Example:

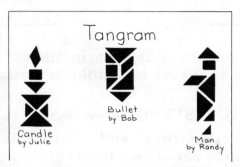

Tangram

Candle
by Julie

Bullet
by Bob

Man
by Randy

26. ONE NUMBER MATH (Grades 6-8)

A. Preparation and Materials: Children will need pencils and paper.

B. Introduction to the Class: Pick any number from 2 through 9. Use **that number only** to make 20 mathematics problems. The answers to your problems will count in order from 1 to 20.

For example, if you picked number 3 you might write:

$$\frac{3}{3} = 1 \qquad \frac{3 + 3}{3} = 2 \qquad \frac{3 \times 3}{3} = 3$$

$$\frac{(3 \times 3) + 3}{3} = 4 \qquad \frac{(3 \times 3) + (3 + 3)}{3} = 5$$

Continue in this same way until the answer to your last problem totals 20.

If you picked the number 7, you might write:

$$\frac{7}{7} = 1 \qquad \frac{7 + 7}{7} = 2 \qquad \frac{7 + 7 + 7}{7} = 3$$

$$\frac{7 + 7 + 7 + 7}{7} = 4 \qquad \frac{(7 \times 7) - (7 + 7)}{7} = 5$$

And you would continue in this same way until the answer to your last problem totals 20.

27. PLACE SIX (Grades 6-8)

A. Preparation and Materials: Draw a grid, like the one shown in the example, on paper. Provide six markers (squares of paper, checkers, beans, etc.).

Example:

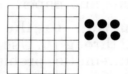

B. Introduction to the Class: Can you place these six markers on six different squares of the playing board so that no two markers are in a line across, down or diagonally?

Solution:

28. PUZZLE CUBES (Grades 6-8)

A. Preparation and Materials: From tagboard cut four cube patterns as shown in the example. Number the faces of each cube as shown. Fold back along dotted lines and glue tabs to form the finished cube shapes.

Example:

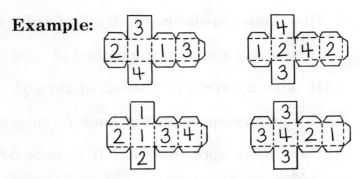

B. Introduction to the Class: I have put four puzzle cubes on the math table. The object of this puzzle is to stack these four cubes, one on top of another, so that each side will show the numbers 1, 2, 3 and 4 in some order. No number will be repeated on any one side of the stack.

Example:

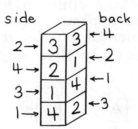

29. HIDDEN SIDE (Grades 6-8)

A. Preparation and Materials: On the board draw the illustration and list the questions given in the example.

Example:

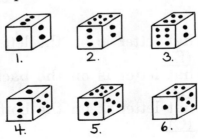

—135—

1. How many spots on the back of cube 1?
   (6)
2. How many spots on the bottom of cube 2 ?
   (1)
3. How many spots on the left of cube 3?
   (5)
4. How many spots on the back of cube 4?
   (4)
5. How many spots on the left of cube 5?
   (2)
6. How many spots on the bottom of cube 6?
   (3)

B. Introduction to the Class: On the board I have drawn six different views of a single dice cube. Study these views carefully. Then, see if you can answer the questions listed.

It won't take children long to discover the spots on two opposite sides add to a total of 7. Then the mental challenge is gone from the puzzle. So on another day try one of these variations.

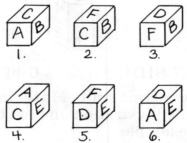

1. What letter is on the left of cube 1?
   (E)
2. What letter is on the back of cube 2?
   (D)
3. What letter is on the bottom of cube 3?
   (C)

4. What letter is on the bottom of cube 4?
  (F)
5. What letter is on the bottom of cube 5?
  (A)
6. What letter is on the left of cube 6?
  (B)

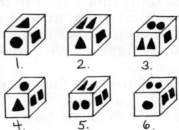

1. What is on the back of cube 1? (▲▲)
2. What is on the bottom of cube 2? (●)
3. What is on the left of cube 3? (■■)
4. What is on the back of cube 4? (●●)
5. What is on the left of cube 5? (■)
6. What is on the bottom of cube 6? (▲)

**30. THE PAINTED CUBES (Grades 6-8)

A. Preparation and Materials: On the board draw the illustration as shown in the example and list the questions.

Example:

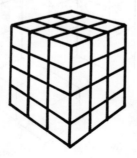

—137—

1. How many separate cubes are needed to build this figure? (36)

2. In the view shown, how many different cubes can you actually see? Remember some single cubes are showing more than one face. (24)

3. Of the total cubes used to make this figure, how many do not show at all in this view? (12)

4. Pretend you have painted all 6 outside faces of this figure blue. How many of the individual cubes:

    a. Have only one face painted blue?
      (10)

    b. Have two faces painted blue?
      (16)

    c. Have three faces painted blue?
      (8)

    d. Have four faces painted blue?
      (0)

    e. Have no faces painted blue?
      (2)

B. Variation: Draw a more complex figure built of individual cubes. List the same kind of questions.

Example:

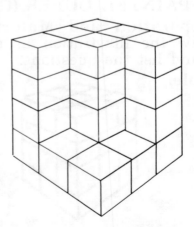

—138—

1. How many separate cubes are needed to build this figure? (27)

2. In the view shown, how many different cubes can you actually see? Remember some single cubes are showing more than one face. (17)

3. Of the total cubes used to make this figure, how many do not show at all in this view? (10)

4. Pretend you have painted all outside faces of this figure blue. How many of the individual cubes:

    a. Have only one face painted blue?
      _(5)_

    b. Have two faces painted blue?
      _(8)_

    c. Have three faces painted blue?
      _(11)_

    d. Have four faces painted blue?
      _(3)_

    e. Have no faces painted blue?
      _(0)_

31. FOUR 4's (Grades 7-8)

A. Preparation and Materials: This is a much more difficult version of One Number Math, described on page 133. Before being able to solve this problem, children must be familiar with square roots and factorials. (The factorial sign is an exclamation point. $4! = 1 \times 2 \times 3 \times 4 = 24$)

B. Introduction to the Class: If you thought One Number Math was fun, try this variation. This is very difficult and presents a genuine challenge.

As in One Number Math, you may use only the number 4 to create 20 equations whose solutions count from 1 to 20. But there is one additional rule. Each equation must contain **exactly four 4's.** You may use any mathematical symbols such as +, −, x, or ÷. You may use decimal points, square roots and factorials, but each problem must contain exactly four 4's, and the answers will count from 1 to 20.

Solution: (Other solutions are possible.)

$$\frac{4+4}{4+4} = 1 \qquad \frac{4 \times 4}{4+4} = 2 \qquad \frac{4+4+4}{4} = 3$$

$$4 \times \left(\frac{4}{4}\right)^4 = 4 \qquad \frac{(4 \times 4) + 4}{4} = 5$$

$$4 + \frac{4+4}{4} = 6 \qquad\qquad 4 + 4 - \frac{4}{4} = 7$$

$$\frac{4! + 4 + 4}{4} = 8 \qquad\qquad 4 + 4 + \frac{4}{4} = 9$$

$$\frac{44-4}{4} = 10 \qquad \frac{4}{.4} + \frac{4}{4} = 11 \qquad \frac{44+4}{4} = 12$$

$$\frac{44}{4} + \sqrt{4} = 13 \qquad\qquad 4 \times 4 - 4 + \sqrt{4} = 14$$

$$(4 \times 4) - \frac{4}{4} = 15 \qquad 4 + 4 + 4 + 4 = 16$$

$$(4 \times 4) + \frac{4}{4} = 17 \qquad 4 \times 4 + 4 - \sqrt{4} = 18$$

$$4! - 4 - \frac{4}{4} = 19 \qquad 4\frac{4}{4} \times 4 = 20$$

32. FOUR ACES, KINGS, QUEENS AND JACKS (Grades 7-8)

A. Preparation and Materials: Take all four aces, kings, queens and jacks from a standard deck of playing cards.

B. Introduction to the Class: Try this puzzle in your free time today. Arrange these cards in four rows of four cards each. No row across, down or corner to corner diagonal may contain two cards of the same suit or value. For example, two jacks may not appear in the same row, nor two hearts.

Solution:

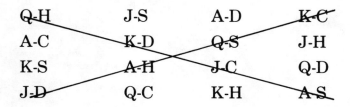

Q-H	J-S	A-D	K-C
A-C	K-D	Q-S	J-H
K-S	A-H	J-C	Q-D
J-D	Q-C	K-H	A-S

CHAPTER IV:
"Are You Really Thinking?"

The test of an open mind: can you look beyond the obvious to discover the unexpected solution for these problems?

INTRODUCTION

A creative mathematician must always have his mind open for surprising twists or unusual angles to a problem — twists or angles other people have overlooked. Let's see if you have that kind of creative mind. Can you look beyond the obvious to discover the tricky "catch" to solve these problems?

(Present only a few of these in any one activity period. If too many are presented at once, it becomes entertainment rather than mental exercise; children want to be given the answers rather than solving the problems themselves. Solve one problem before going on to the next. If no one can think of an answer, let them "sleep on it" and bring in a solution on the following day.)

1. THE RAPID READER: A man found he could read 80 pages of a book in 80 minutes when he wore his red shirt. But when he wore his blue shirt, it took him an hour and 20 minutes to read the same number of pages. How could this be true?

SOLUTION: The shirt, of course, has nothing to do with it. Eighty minutes and an hour and 20 minutes are the same length of time.

2. PUZZLING PAGES: Sue needed magazine pictures for a science project. She cut out pages 20, 21, 97, 98 and 104. How many sheets of paper did she cut from the magazine?

SOLUTION: 4 sheets. Page 97 is on the back of page 98.

3. COMMON CENTS: (Write this puzzle on the board.) Why are 1975 pennies worth almost $20.00?

SOLUTION: 1,975 pennies = $19.75, which is almost $20.00.

4. THE STRANGE SURVEY: A government survey was taken in the city of Mathville. The survey team measured the length of every resident's right foot and gave each person a test in mathematical ability. The survey proved a direct correlation between foot size and mathematical ability. How could this be so?

SOLUTION: The survey included **everyone** in town. Naturally babies and young children (small foot size) would have less mathematical ability than adults (large foot size).

5. HUNGRY HARRY: Harry had 10 candies and ate all but two. How many candies did he have left?

SOLUTION: Two. (He did not **eat** two. He ate **all but two.**)

6. THE CASE OF THE RACING HEIRS: A man died, leaving two sons. His will decreed his sons should have a horseback race from their barn to a bridge exactly 3 miles away. The man's entire estate would go to the son whose horse reached the bridge **last.**

The sons immediately mounted the two horses and raced toward the bridge at breakneck speed.

Why would the sons ride so fast, when the winner was to be the one whose horse reached the bridge last?

SOLUTION: Each son rode his brother's horse. His own horse would arrive last if he could get his brother's horse to the bridge first.

7. THINK QUICK: 5 x 3 is 15, and 4 x 4 is 16. Can you tell me two whole numbers that make **17** when multiplied together?

SOLUTION: 17 x 1 = 17.

8. END OF THE LINE: Let's pretend you are the driver of a city bus. At the first stop 11 passengers get on. At the next stop 3 get on and 4 get off. At the next stop 7 get on and 2 get off. At the next stop 9 get on and 2 get off. At the next stop 4 get on and 5 get off.

Now, can you tell me, does the driver of the bus have red hair?

SOLUTION: You are the driver. Do you have red hair?

9. ORDER, PLEASE: I am going to write the numerals 0 through 9 on the board in a definite order. Can you figure out the logic for this order of numbers? Write the following number sequence on the board:

8 - 5 - 4 - 9 - 1 - 7 - 6 - 3 - 2 - 0

SOLUTION: They are in alphabetical order according to the written word for each numeral.

10. THE PUZZLING PARROT: "The parrot is a fantastic bird," said the pet store salesman. "He can give the correct answer to absolutely every mathematical problem he hears."

The customer was so impressed he bought the parrot. But when he got it home, the bird wouldn't answer a single problem.

If what the salesman said was true, why wouldn't the bird answer the problems?

SOLUTION: The parrot was deaf.

11. COMMON ERROR: What common English word is pronounced wrong by every mathematician in America?

SOLUTION: The word "wrong."

12. THE PILL PUZZLE: Your doctor gives you three pills and instructs you to take one every half hour. How long will your pills last?

SOLUTION: One hour. You take one now, one a half hour from now (total time lapse: ½ hour), and one a half hour later (total time lapse: 1 hour).

13. WHEEL WIZARDRY: A bicycle wheel has 35 spokes. How many spaces does it have between spokes?

SOLUTION: 35.

14. HIGH FINANCE: If you took 5 coins from a piggy bank containing 17 coins, how many coins would you have?

SOLUTION: The 5 you took.

15. DISSOLVING DILEMMA: A chemist rushed from his laboratory carrying a flask of steaming, purple liquid. "Look what I've invented," he said proudly to a fellow chemist. "The liquid in this bottle is such a powerful acid that it will dissolve anything it touches."

"You're wrong," said the second chemist immediately.

How did the second chemist know that the first was wrong?

SOLUTION: If the liquid would dissolve anything it touched, it would have dissolved the flask in which it was being carried.

16. ANOTHER STRANGE SURVEY: A recent survey in the town of Blue Rock found a surprising number of residents who could not speak the English language. These findings startled the townspeople for there were no foreigners at all living in the town.

But the survey was correct. How could this be true?

SOLUTION: The non-English speaking residents were babies, too young to speak **any** language.

17. DOG-GONE CLEVER: A dog has a 10 foot chain attached to his collar. He sees a bone 11 feet from where he is standing. He looks and looks at the bone, and finally figures out a way to get the bone. How does he do it?

SOLUTION: He simply walks over and takes it. One end of the chain is attached to his collar, but the other end is attached to nothing.

18. THE SOCK PUZZLE: You have 10 blue socks and 10 red socks in your dresser drawer. The room is dark so you cannot see into the drawer. How many socks would you need to take from the drawer to be sure you had a pair that matched.

SOLUTION: Only 3. Two of those three would be the same color.

19. THE SOCK PUZZLE AGAIN: Suppose you wanted a pair of **red** socks from the drawer previously described. How many socks would you have to take from the drawer to guarantee getting two red ones?

SOLUTION: 12 socks. It would be possible to take out 10 blue socks in a row, but the next two would have to be red.

20. THE CAGEY CASHIER: A lady walked up to the lunch counter to pay her bill. The cashier noticed the lady had carefully drawn a circle, square and triangle along the left hand side of the bill. The cashier immediately looked up and said, "How long have you been a policewoman?"

How did the cashier know the lady was a policewoman?

SOLUTION: The lady was wearing her uniform.

21. A WEIGHTY PROBLEM: Which weighs more, a pound of lead or a pound of feathers?

SOLUTION: They are equal.

22. THE COIN CAPER: I have two coins that total 15 cents. One coin is not a dime. What are my two coins?

SOLUTION: A dime and a nickle. (Yes, I said **one** coin **was** not a dime. But the **other** coin is.)

23. A KNOTTY PROBLEM: A man attached a knotted rope to the side of his boat. The lowest knot was barely underwater. The remaining knots were all 5 inches apart.

When the tide rises 3 feet, how many knots will be underwater?

SOLUTION: Still only one. The boat floats and will rise with the tide. The rope is attached to the boat and will rise also.

24. DIVIDE AND CONQUER: Divide 10 by ½ and add 10. What is your answer.

SOLUTION: 30. (Did you say 15? Shame on you. I said divide **by** ½ not divide it in half.)

25. THE STRANGE SHOPPER: A man was shopping for a certain item in a hardware store. "How much for 6?" he asked the clerk. "50 cents," said the clerk. "How much for 60?" asked the man. "$1.00," answered the clerk. "I'll take 600," said the man. "That will be $1.50," said the clerk. What was the man buying.

SOLUTION: House numbers.

26. ALARMING PUZZLE: A mathematician went to bed at 9:00 one night and set his alarm to awaken him at 10:00 the next morning. When the alarm went off, how long had he slept?

SOLUTION: One hour.

27. GROOVY: How many grooves are there on one side of a 12-inch LP record?

SOLUTION: Only one. It begins on the outer edge and spirals in towards the center.

28. BOILING EGGS: If it takes 5 minutes to hard-boil one egg, how long does it take to hard-boil 5 eggs?

SOLUTION: Just 5 minutes. All the eggs are in one kettle.

29. THE BICYCLING BROTHERS: Two brothers were bicycling down the road. The brother in front said to the other, "I'll bet you a dollar you can't pass me."

The brother in the rear was older, bigger and stronger, but he replied, "You know I could never win that bet."

Why was this true?

SOLUTION: The boys were riding a bicycle built for two.

30. THINK FAST: Name the months that have 30 days.

SOLUTION: All the months except February.

31. TWIN TROUBLE: A boy saw two girls in the park. They looked exactly alike, so he asked, "Are you girls twins?"

"No, we're not twins," said one girl.

"That's right," said the other. "We have the same parents, and we were born on the same day of the year, but we're not twins."

How could this be so?

SOLUTION: The girls were two from a set of triplets.

32. TIGHT SQUEEZE: Two boys found a tunnel through a giant rock. The tunnel was just barely big enough for one boy to squeeze in and crawl from one entrance to the other.

If each boy starts at an opposite end of the tunnel, could each crawl through the full length of the tunnel and come out on the other side?

SOLUTION: Certainly. One boy crawls through in one direction. **After he comes out,** the other boy crawls through in the opposite direction.

33. SOMETHING TO CROW ABOUT: Twelve crows were sitting in an oak tree. A farmer shot into the tree, killing one crow. How many crows were left?

SOLUTION: None. The remaining crows, frightened by the gunshot, would fly away.

34. NORTH IS SOUTH?: A man was in his car headed due north. The road was perfectly straight. After driving one mile and making no turns at all, the driver discovered he was exactly one mile **south** of his starting place. How could this be so?

SOLUTION: The man drove his car backwards.

35. THE ELEVATOR PUZZLE: Mr. Gates worked hard each day, and by the time he reached home he was so tired he could hardly put one foot in front of the other.

Yet every night he got into the elevator, pushed the 7th floor button, got off at the 7th floor and walked the remaining three flights of stairs to his 10th floor apartment. Why did he do this?

SOLUTION: Mr. Gates was a midget, and could reach only as high as the 7th floor button.

36. RELATIVELY SPEAKING:

Brothers and sisters have I none,
But this man's father is my father's son. What is my relationship to this man?

SOLUTION: I am his father. Logic:
GIVEN: [This man's **father**] = [My father's **son**]
KNOWN: I am my father's son.
THEREFORE: I am this man's father.

37. MORE RELATIVES: Two farmers were plowing a field. The tall farmer was the short farmer's father, but the short farmer was not the tall farmer's son. How could this be so?

SOLUTION: The short farmer was the tall farmer's daughter.

38. TEN DOLLARS: Which is worth more, a new ten dollar bill or an old one?

SOLUTION: The new ten dollar bill is worth 9 dollars more than the old **ones.**

39. WOODCUTTERS: If it takes 6 men 6 minutes to cut down 6 trees, how long does it take 5 men to cut down 5 trees?

SOLUTION: 6 minutes.

40. PLANE PUZZLE: An airplane left New York headed for Los Angeles at a speed of 500 miles per hour. Half an hour later another plane left Los Angeles headed for New York at 600 miles per hour. When the two planes meet, which is closest to Los Angeles?

SOLUTION: When the planes meet, they are in the same location; therefore, each is the same distance from Los Angeles.

41. COUNTING TREES: On his way to school Bill counted 37 trees along the left side of the street. On the way home he counted 37 trees on the right side of the street. How many trees did Bill count in all?

SOLUTION: 37 trees. The trees on the left side going to school are the same ones on the right side as he comes home.

42. JUDY'S GRANDMOTHER: Judy says her father is 45 years old and her grandmother is 53. Her father couldn't possibly have been born when his mother was only 8 years old. So how could this be so?

SOLUTION: Judy is speaking of her grandmother on her **mother's** side. Her mother is only 35 and was born when Judy's grandmother was 18.

CHAPTER V:
"Geometric Art"

Mathematical concepts applied to the field of art produce fascinating results.

1. PUNCHED SHADOW DESIGNS
(Grades 4-8)

A. Preparation and Materials: Paper, pencils, rulers, compasses, nails and a soft towel.

B. Introduction to the Class: Draw a geometric design on paper. Use a ruler or compass if needed. Do not take time to erase any errors because this design will become the **back** of your finished picture.

When your design is complete, place it on several layers of soft towels. Use a nail to punch holes at regular intervals along each line of your design.

Turn your paper over. The rows of raised punch marks on the reverse side of the paper forms the design. When displayed on the wall the raised texture casts an interesting play of light and shadow.

Example:

Working side (back) of picture: holes punched along lines of design.

Reverse side (front) of picture: punches form textured design.

*2. KALEIDOSCOPE DESIGNS (Grades 4-8)

A. Preparation and Materials: Each child will need drawing paper, compass, protractor, ruler, pencil and crayons, felt pens or colored pencils for coloring the designs.

*This activity is available in Challenge Volume I of the **Spice**™ Duplicating Masters.

B. Introduction to the Class:

Step 1: Draw a circle. Draw lines passing through the center of the circle to divide the circle in half vertically and horizontally.

Step 2: Color a design in the center of the circle. Draw identical designs on each of the four spokes of the circle.

Step 3: On the circle's circumference mark the mid-points between each existing spoke. Draw lines to connect each of these four points with the center.

Step 4: Color identical patterns along each of these four new spokes.

.** 3. GRAPH PAPER ART (Grades 4-8)

A. Preparation and Materials: Graph paper with four squares per inch (a smaller grid pattern can be used, but designs become very tiny and insignificant), pencils, ruler, compasses, felt pens or colored pencils, etc.

B. Introduction to the Class:
Old Fashioned Samplers: Have you ever seen cross stitch embroidery patterns? In a cross stitch sampler a picture is made from many little embroidered X's. You can make the same kind of picture using graph paper.

First, use a pencil to lightly sketch the outline of the design on your graph paper. In each **full square** enclosed in your design make an X. If only part of a square is within the border of your design, leave that square empty. The edges of the design will be like little stair steps.

Use colored pencils or felt pens for these X's. Use all one color or a variety of colors as you prefer. Crayons do not work well because they make too wide a line. After your coloring is done, erase all penciled guide lines that show.

Example:

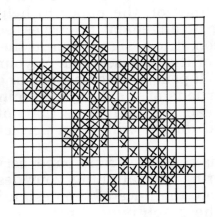

*This activity is available in Challenge Volume I of the Spice™ Duplicating Masters.
**This activity is available in Challenge Volume II of the Spice™ Duplicating Masters.

Block Design: First, lightly sketch the design on your graph paper using a pencil. Solidly color each **full square** enclosed in your design. If only part of a square is within the border of your design, leave that square empty.

As in the cross stitch designs, use all one color or many colors. Erase all penciled guide lines that show. Your design will show in solid blocks of color.

Example:

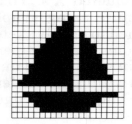

Graph Designs: Using the graph paper lines as guides, divide the paper in half vertically and horizontally. In each of these four sections, use the lines on the paper to help you make a pleasing arrangement of circles, squares or any other geometric designs. Repeat the same pattern in each of the four sections of your paper.

Use felt pens, colored pencils or crayons to color in portions of the design, keeping colors identical in each of the four sections of your design. The total design will be in perfect symmetrical balance.

In this case, the graph paper serves only to aid in centering and aligning your design. In coloring, follow the lines of the design itself and disregard the graph paper lines.

Example:

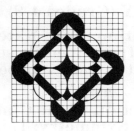

Tessellation: Color squares, parts of squares or graph lines to form mosaic patterns. Make one large design on the page or make a repeated pattern of small designs. (See also Symmetry in Repeated Patterns, page 165.)

Example:

Or, make a small, basic pattern. See how many different ways you can combine that one shape only to form a variety of overall patterns.

Example: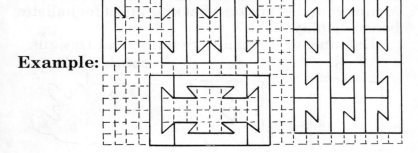

4. SYMMETRY (Grades 4-8)

A. Preparation and Materials: Children will need colored paper, pencils, scissors and paste, etc., as directed in each numbered section that follows.

B. Introduction to the Class:
Vertical Symmetry: Think of a mirror set upright along the vertical center line of a design. If each vertical half of the design is an exact reflection of the other, the design is said to have vertical symmetry.

In our alphabet the letters A, H, M, O, T, V, W, X and Y have vertical symmetry.

Example:

Fold a piece of paper in half vertically. Cut a shape from the folded paper. When you open your paper, you will find the shape you have cut has vertical symmetry. Each half of the design is an exact reflection of the other. Try cutting some shapes of this kind. The perfect balance of the two sides makes the designs artistically pleasing. Mount your designs on a backing sheet for bulletin board display.

Example of Vertically Symmetrical Designs:

fold   cut   completed design

Horizontal Symmetry: Now, think of a mirror set upright along the horizontal center line of a design. If each horizontal half of the design is an exact reflection of the other, the design has horizontal symmetry.

In our alphabet the upper case letters C, D, E, H, I, O and X have horizontal symmetry.

Example:

----C-D-E-H-I-O-X-

Fold a piece of paper in half horizontally. Cut a shape from the folded paper. Open the paper to find a design with horizontal symmetry. Notice the perfect balance of the design in the top and bottom half of the sheet. Mount your design for bulletin board display.

Example of Horizontally Symmetrical Designs:

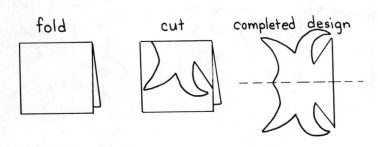

fold   cut   completed design

Vertical and Horizontal Symmetry:
Think of two lines cutting a design in half both vertically and horizontally. If each of these four sections is an exact reflection of the sections it touches, the design has both vertical and horizontal symmetry.

In our alphabet the upper case letters H, O and X have both vertical and horizontal symmetry.

Example:

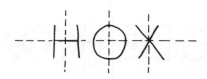

To cut a design with both vertical and horizontal symmetry fold the paper in half both vertically and horizontally. Cut through all four thicknesses to create a design. Open the folds to find a perfectly symmetrical shape.

Example:

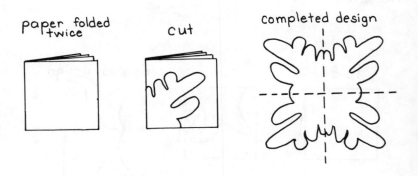

paper folded twice

cut

completed design

Symmetry in Repeated Patterns:
Knowledge of symmetry can help you arrange a repeated pattern in an artistically pleasing way.

a. Translation: Each design maintains its original position throughout the pattern.

Example:

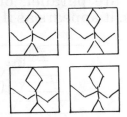

b. Rotation: Turn or rotate a design a quarter turn or a half turn each time it is repeated.

Example:

Quarter Turn   Half Turn

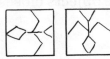   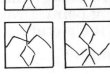

c. Reflection: Turn each shape so it is an exact reflection of each shape it touches.

Example:

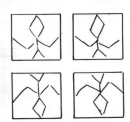

5. VANISHING POINT (Grades 4-8)

A. Preparation and Materials: Children will need pencils, paper and a ruler.

B. Introduction to the Class: You know that railroad tracks are parallel, but when you look down the tracks, it appears as if the two rails meet in the distance.

Example:

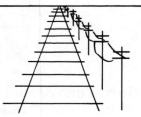

Drawing pictures to make objects grow smaller as they recede into the distance is really quite easy to do. First, draw the horizon line. This is the line where the sky and earth meet. Make one dot anywhere along that line. The dot will be your vanishing point. Draw in the objects closest to the front of the picture.

Example:

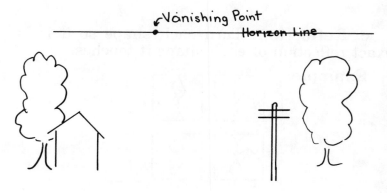

Draw lines to connect the top and base of each object with the vanishing point. Draw in the remainder of the picture fitting the size within the lines going to the vanishing point. The objects thus grow smaller as they recede into the distance.

Example:

Vanishing Point

6. GEOMETRY IN FAMOUS ART
(Grades 4-8)

A. Preparation and Materials: Collect books or prints illustrating art by such famous artists as Mondrian, Klee, Kandinsky, Feininger and/or Matisse. Also have illustrations of art by the famous cubists Picasso, Barque, Juan Gris and/or Albert Gleizes. Each of these artists has famous paintings composed totally of basic geometric shapes.

B. Introduction to the Class: You know that geometry is an extremely useful branch of mathematics, but look what these artists have done using geometry in art.

(Show the illustrations and discuss the shapes used in each painting. This discussion can lead to further study by individual students. Some with artistic ability may wish to create their own geometric art for classroom display. Others might bring in prints showing more examples of geometry used in art form. Some may wish to give oral or written reports on a particular artist or his work. You may wish to set aside a section of the bulletin board to display prints and reports illustrating geometry in art.)

7. COMPASS AND RULER ART
(Grades 4-8)

A. Preparation and Materials: Children will need paper, pencils, compasses, rulers and colored pencils, crayons or felt pens.

B. Introduction to the Class: Probably the best way to explain how to make these designs is simply to show you some examples and then let you experiment making designs of your own. Use your compass to make circles or arcs and your rulers to make shapes with straight edges. Experiment on scratch paper until you get a design you like. Copy that design onto good paper and color it.

Example:

A. Preparation and Materials: White or colored paper, colored felt pens, rulers, pencils, compasses, etc.

B. Introduction to the Class: Here's a method for making geometric designs that seem to spin or vibrate. The colors you use for your design are important and so is the composition of the design itself.

Basic Working Method: The basic design consists of parallel stripes of alternating colors and dividing bars.

Take a sheet of paper and lightly pencil in dots every quarter inch along the left and right margins. (Lines may be spaced from one-eighth to one-half inch apart in proportion to the size of the finished work.) Use these dots as guides to draw horizontal, parallel lines across the page.

Example:

Next, place five or six upright dividing bars anywhere on the paper. The bars can be vertical or slanted. Space some close together and some further apart.

Example:

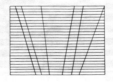

*This activity is available in Challenge Volume I of the **Spice**™ Duplicating Masters.
This activity is available in Challenge Volume II of the **Spice™ Duplicating Masters.

Now, the design is ready to color. Choose two colors for your design. Using bright red and bright green produces a vibrating effect that makes you downright dizzy. A combination of bright blue and bright green, or yellow/orange and bright green is also effective. But for this first practice drawing we'll use black and white.

Begin coloring in the left section between the left edge of the paper and the first dividing bar. In that section color every other stripe black.

Example:

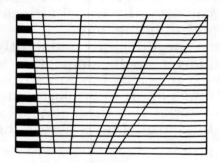

Between the next set of dividing bars color every other stripe black, but alternate colors with the previous section. If a stripe was white in the previous section, color it black in this section. Continue in this way to complete the design. Erase any penciled guide lines that show.

Example:

Dividing Bar Variations: Variations in the shape of the dividing bars will produce totally new effects.

Example:

Curved Dividing Bars   Zig-Zag Dividing Bars

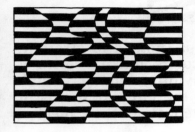 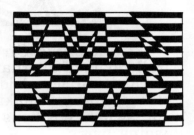

Zig-Zags: First, put in the dividing bars. Mark off dots at quarter inch intervals along the right and left margins of the paper and along each dividing bar. (See Example 1.) Connect pairs of dots in a zig-zag pattern across the sheet. (See Example 2.)

Example 1   Example 2

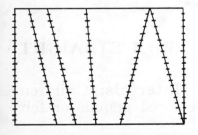 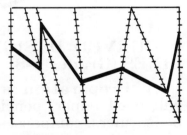

Using the dots as guides, lightly pencil in lines in each section parallel to the first drawn. Color every other stripe black switching positions of the colors each time a dividing bar is met.

Example:

****Concentric Circles:** Instead of straight, parallel bars of alternate colors, make a series of concentric circles. Add dividing bars as usual. Color the design making the circular bands of color switch positions each time they meet a dividing bar.

Example:

****9. CURVED DESIGNS FROM STRAIGHT LINES (Grades 4-8)**

A. Preparation and Materials: Children will need paper, pencils, colored pencils or felt pens, rulers, compasses, etc.

B. Introduction to the Class:
Basic Working Method: I will show you how to make curved designs using straight lines only.

First, draw two lines of equal length. Use your ruler so the lines will be straight. Make a dot every quarter inch along each line. Mark these dots very lightly so they will not show when your design is complete. (Dots can be spaced one-eighth to one-half inch apart depending on the size of the finished work.)

Example:

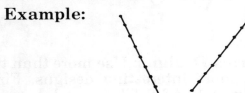

Think of the dots on the left line as being numbered in order from top to bottom. On the right think of the dots being numbered in order from bottom to top. Use your ruler to draw a line connecting the two #1 dots. Then connect the two #2 dots, the two #3 dots and so on. Use all one color for your lines or change colors every few lines.

Example:

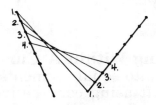

Continue in this way until each pair of dots has been connected. See the graceful curved design produced by the intersection of perfectly straight lines?

Example:

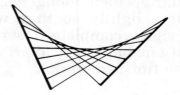

Line Series Designs: Use more than two lines to create more interesting designs. First, work between lines 1 and 2. Then, work between lines 2 and 3 in the very same way, and so on to complete the design.

Example:

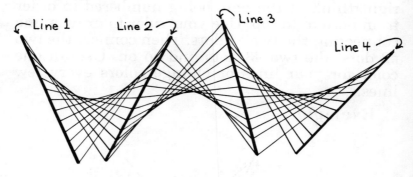

Working Within A Curve: Draw an arc and mark it into equal segments. Begin numbering on the left using consecutive numbers until you reach the center of the arc. Number the center dot 1, the next dot 2, and so on until you reach the right end of the arc.

Draw a straight line to connect the two dots numbered 1, then the two dots numbered 2 and so on to complete the design.

Example:

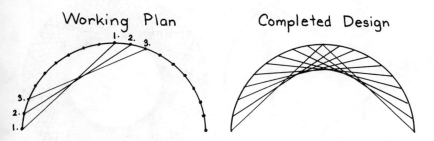

Working Plan   Completed Design

Combined Curve Designs: Combine two or more curves to create more intricate designs.

Example:

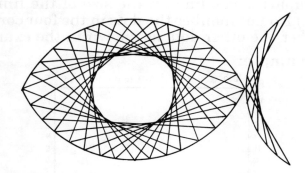

**Working Within A Circle: Draw a circle. Use a protractor to mark dots every 5⁰ on the circumference. Beginning anywhere on the circumference, number consecutive dots from 1 to 24. Number the next dot 1 and continue numbering successive dots through 24. Then begin at 1 and number through 24 again. You will have three sets of 24 dots, then connect each set of number 2 dots and so on to complete the figure. Notice these straight lines create a central circle.

*This activity is available in Challenge Volume I of the **Spice**™ Duplicating Masters.
This activity is available in Challenge Volume II of the **Spice™ Duplicating Masters.

Example:

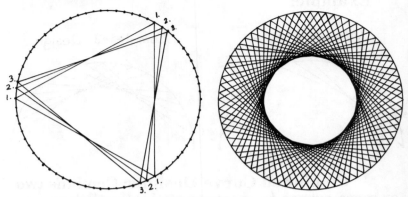

*Working Within A Square: Draw a square. Mark dots at equal intervals along each edge. (Dots can be spaces one-eighth to one-half inch apart depending on the size of the finished work.) Do not number the dots in the four corners. Number the other dots as shown in the example.

Example:

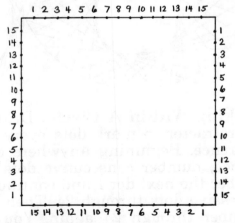

Connect each dot with a dot of the same number on each of the two adjoining sides. (See Figure 1.) Continue in this way to complete the design. (See Figure 2.)

—176—

Figure 1

Figure 2

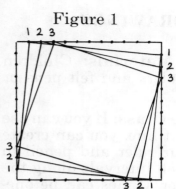

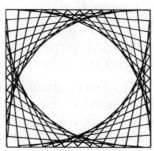

Free Form Designs: Once you have mastered the basic working method, you can begin creating your own line designs. Try this free form design. Draw a circle inside a quadrilateral figure. Use pencil so these shapes can be erased when the design is complete. Draw a series of lines tangent to the circle and stopping at the edges of the quadrilateral figure. These lines do not need to be evenly spaced. The lines will outline the shape of both the circle and the quadrilateral shape.

Example:

Working Method   Completed Design

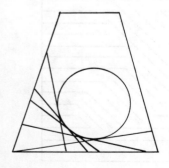

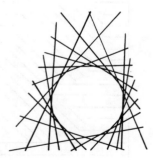

10. PARALLEL LINE DRAWINGS
(Grades 6-8)

A. Preparation and Materials: Children will need paper, pencils, rulers and felt pens or colored pencils.

B. Introduction to the Class: If you can use a ruler and draw a straight line, you can create artistic drawings. Use your ruler and pencil to mark off dots at quarter inch intervals along the outside edges of your paper. (Lines can be one-eighth to one-half inch apart depending on the size of paper used and the width of the pen or crayon used.) Use these dots as guides for drawing parallel lines for the background. Parallel lines running in a new direction form the design. After completing the parallel colored lines, erase all penciled guide lines that show.

Free Form Designs: The interior free form pattern is defined by parallel lines tipped at a slight angle from the horizontal background bars. Erase all penciled guide lines that show.

Example:

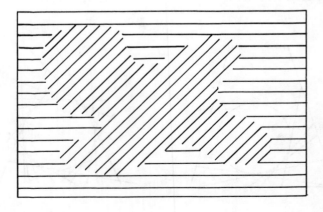

Repeat Patterns: Alternating vertical and horizontal lines add variation to a repeated design. After completing the work, erase all penciled guide lines that show.

Example:

Overlapping Shapes: Overlapping shapes can create unusual effects. Step 1: Draw two overlapping circles. Step 2: Draw horizontal lines for the background and center of the design. Step 3: Draw vertical lines in the remaining areas. Erase all penciled guide lines that show.

Example:

Step 1   **Step 2**   **Step 3**

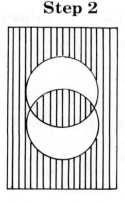

11. GEOMETRIC MOBILE (Grades 6-8)

A. Preparation and Materials: Children will need medium weight paper, glue, pencils, rulers and scissors. Additional supplies may be used for decorations.

Suspend the finished shapes on threads hung from wooden dowel lengths or lengths of heavy wire to make the completed mobiles.

Example:

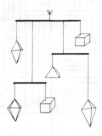

B. Directions:

Tetrahedron (four equal sides): Lay out a pattern of four equilateral triangles as shown in the illustration. Add tabs for joining the form together.

Fold up along all dotted lines and glue tabs in place to assemble.

Closed Tetrahedron

Pattern   Assembled Form

If desired, decorations can be added before the shape is assembled. Designs can be drawn on with felt pens or crayons. Colored wrapping paper may be pasted over the pattern. Or, the inside of each plane surface may be cut out to make an open design.

Open Tetrahedron

Pattern   Assembled Form

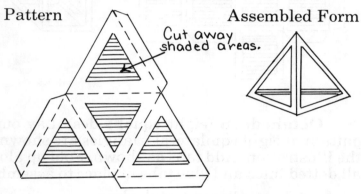

Cut away shaded areas.

Cube (six equal sides): Lay out a pattern of six squares as shown in the illustration. Add tabs as shown. Fold up along all dotted lines and glue tabs in place to assemble.

Closed Cube

Pattern   Assembled Form

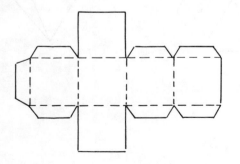
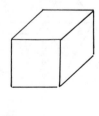

Open Cube

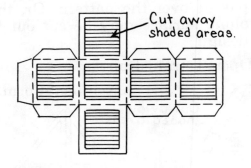

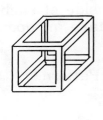

Pattern

Assembled Form

Octahedron (eight equal sides): Lay out a pattern of eight equilateral triangles are shown in the illustration. Add tabs as shown. Fold up along all dotted lines and glue tabs in place to assemble.

Closed Octahedron

Pattern

Assembled Form

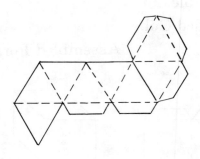

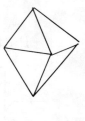

Open Octahedron

Pattern   Assembled Form

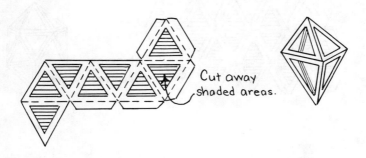

Cut away shaded areas.

Icosahedron (twenty equal sides): Lay out a pattern of twenty equilateral triangles as shown in the illustration. Add tabs as shown. Fold up along all dotted lines and glue tabs in place to assemble.

Closed Icosahedron

Pattern   Assembled Form

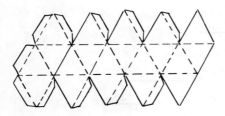

Open Icosahedron

Pattern   Assembled Form

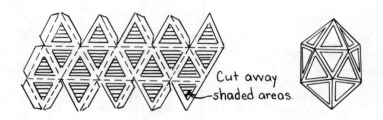

Cut away shaded areas.

12. HANGING IN BALANCE (Grades 6-8)

A. Preparation and Materials: You will need slender dowel rods, a single-edged razor blade for cutting the dowel lengths to size, thread and a set of **identical** heavy weights. Washers, nuts, bolts, lead sinkers, etc., can be used for weights, but all weights used in a single mobile **must be identical.**

B. Directions: Cut dowel lengths, tie on weights and assemble as shown in the illustrations for Pattern 1 and Pattern 2. These mobiles will hang in perfect balance.

Pattern 1:

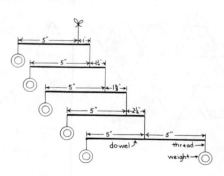

Pattern 2:

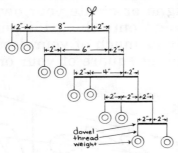

**13. DISTORTIONS (Grades 6-8)

A. Preparation and Materials: Children will need paper, pencils and rulers.

B. Introduction to the Class: You have learned how to copy a design from one grid onto a larger or smaller grid in order to enlarge or reduce the size of that design. (See Enlarging or Reducing Patterns, page 192.)

Today, we're going to copy designs from a square grid onto a grid of a totally different shape in order to create interesting distortions of that design.

First, draw a square grid. Draw a design on that grid.

Example:

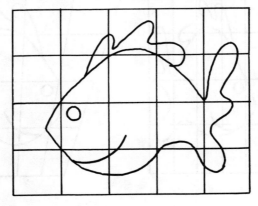

—185—

This activity is available in Challenge Volume II of the **Spice™ Duplicating Masters.

Now, draw grid variations. Try these illustrated designs or create your own variations. Copy your design onto the grids putting in each section exactly as much of your design as was in the corresponding square of your original grid.

Example:

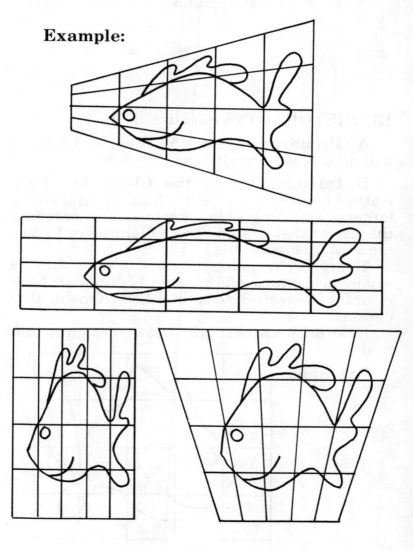

CHAPTER VI:
"Miscellaneous"

Additional activities useful in the intermediate mathematics program.

1. DRAWING BASIC SHAPES USING NO SPECIAL TOOLS (Grades 4-8)

A. Preparation and Materials: Pencils, paper, string and pins.

B. Introduction to the Class: Here are quick ways to draw some basic geometric shapes using no special equipment.

Straight Line: You want to draw a straight line, but you have no ruler. Here's a quick way. Simply fold any piece of paper. The folded edge forms a perfect straight line. Use that edge to guide your drawing.

Example:

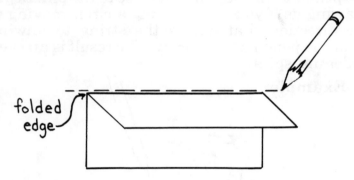

folded edge

Circle: You need to draw a circle and you have no compass? Just tie a string around your pencil. Push a pin through the other end of the string. Stick the pin upright in a piece of paper. (A magazine placed under the paper helps hold the pin securely.) Hold your pencil upright and pull gently outward until the string is taut. Keep the string taut as you rotate the pencil around the pin and you'll draw a perfect circle. Lengthen or shorten the string to draw bigger or smaller circles.

Example:

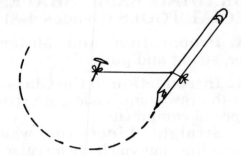

Spiral: Use the pin, string and pencil as described for the circle but wind the string around the pencil until the pencil is close to the pin. Begin drawing as if you were making a circle moving in the direction that allows the string to unwind from the pencil as you draw. The result is an ever-widening spiral.

Example:

Ellipse: Put two pins into your paper about 2 inches apart. Tie a loop of string about 8 inches long. Lay the loop over the pins. Put the pencil into the loop and pull gently outward until the string is taut. Keep the string taut as you rotate the pencil around the pins. Lengthen or shorten the string to enlarge or reduce the size of the ellipse. See what happens when you place the two pins closer together or further apart.

Example:

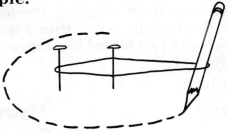

Fold a Square: Take a rectangle of paper. Fold Edge A down to perfectly align with Edge B. Cut away the length of paper extending beyond the two layers of paper. Open the folded section and you will see a perfect square.

Example:

Initial Rectangle

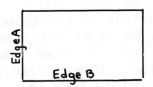

Fold & Cut   Completed Square

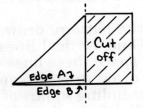

Fold a Right Angle: Fold any shape of paper in half. The fold line forms a perfectly straight line. (See Figure 1.) Fold this straight folded edge over itself aligning the top layer exactly over the bottom layer. (See Figure 2.) The corner where the two folded lines meet is a perfect right angle.

Example:

Figure 1   Figure 2

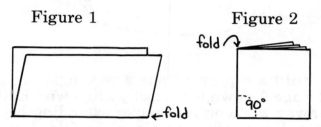

A. Preparation and Materials: Paper, pencils and rulers.

B. Introduction to the Class: Suppose you find a pattern for something you would like to make, but you want to make it bigger or smaller than the pattern shows. Here is one way to change the size of the pattern.

First, draw vertical and horizontal lines over the top of the pattern. Space the lines about one-half inch apart.

If you want to enlarge the pattern, draw the same number of horizontal and vertical lines on another sheet of paper, but space these lines one inch apart. On this paper draw in each square exactly what is in the corresponding square of the pattern. Your new picture will be exactly twice as big as the original pattern.

*This activity is available in Challenge Volume I of the Spice™ Duplicating Masters.

Example: Enlarging a Pattern

Original Pattern   Enlarged Pattern

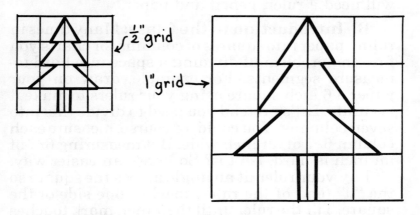

If you want to reduce a pattern, first, draw the same horizontal and vertical lines on the original pattern spacing the lines one-half inch apart. On another sheet of paper draw a grid with vertical and horizontal lines spaced one-quarter inch apart. On this paper draw in each square exactly what is in the corresponding square of the pattern. Your new picture will be exactly half the size of the original pattern.

Example: Reducing a Pattern

Original Pattern   Reduced Pattern

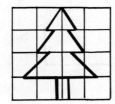

3. HOW TO DIVIDE SPACES (Grades 4-8)

A. Preparation and Materials: Each child will need a ruler, pencil and paper.

B. Introduction to the Class: Many times in ruling paper into squares of columns for charts you face the problem of dividing a space into hard-to-measure segments. For example, draw on your paper a 6 inch square using your rulers to make it accurate. Let's pretend you need to divide this into seven columns. You could, of course, measure each column 6/7 of an inch wide. But measuring 6/7 of an inch is hard, isn't it? So here's an easier way.

Lay your ruler at an angle across the square so the "0" (end of the ruler) touches one side of the square. Tilt the ruler until the 7 inch mark touches the opposite side. (Demonstrate. See example.)

Example:

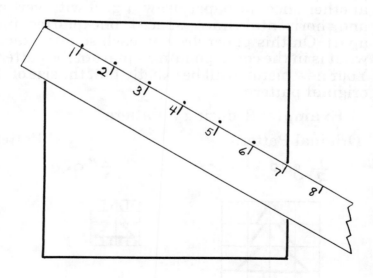

Lightly mark a dot on your paper by each inch mark on the ruler. (Dots are shown in the example.)

Now tip the ruler the opposite way, and make a second series of dots.

Example:

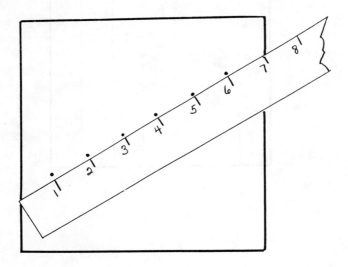

Use your ruler to line up each pair of dots that are directly above each other. Draw a line from the top of the square to the bottom, passing through each pair of dots. You have now divided a 6 inch square into 7 equal columns.

Example:

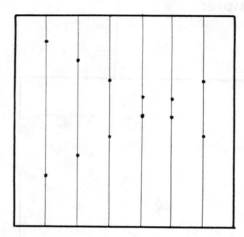

Now draw a five inch square. Can you divide it into six equal columns? (Continue in this way until you are sure the children understand the process.)

(In a future lesson you may wish to add further explanations. If, for example, you wish to divide a 3 inch square into 7 segments, tip the ruler so the "0" and "3½" touch opposite sides of the square. Place a dot by each half inch mark.

To divide a seven inch square into four equal segments, tip the ruler until the "0" and "8" touch opposite sides. Place a dot by the 2, 4, and 6 inch marks, thus cutting the space into two inch segments.)

4. TYING SHAPES (Grades 4-8)

A. Preparation and Materials: You will need strips of paper (the width of the paper determines the length of a side of the finished shape) and scissors.

B. Introduction to the Class:

Pentagon: Take a strip of paper. Label the ends A and B.

Bring end A across the strip at a right angle.

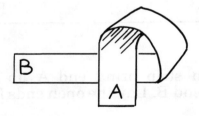

Tuck end B through the loop.

Pull ends A and B to tighten the knot. Flatten the folds and cut off the extra length.

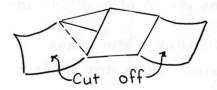

Hexagon: Use two strips of paper. Label the ends of each strip A and B.

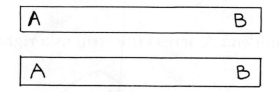

With each strip bring end A up and lay it overlapping end B. Lay the open ends facing each other.

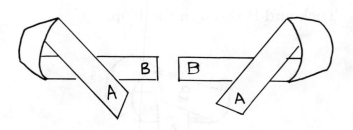

Slip the open ends of each strip through the loop of the other strip.

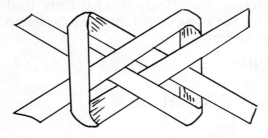

Pull the knot tight, flatten the folds and cut off extra length.

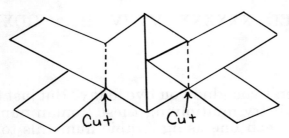

5. MULTIPLYING ROMAN NUMERALS (Grades 4-8)

A. Preparation and Materials: Put on the board the problems shown in the example.

Example:

$$
\begin{array}{r}
\text{XXVII} \\
\text{x XVI} \\
\hline
\end{array}
\qquad
\begin{array}{r}
27 \\
\text{x 16} \\
\hline
\end{array}
$$

B. Introduction to the Class: Before you ever complain about multiplication again, take a look at what the Roman children had to do. (Proceed to work the problem as shown explaining each step as you go.)

Example:

XXVII		27
XVI		x 16
XX V II		162
LL XX VVV	vs.	27
CC L XX		432

CC LLL XXXXXX VVVV II = CDXXXII

Aren't you glad you live today? But just for the fun of it try multiplying these Roman numerals. Check each one using Arabic numerals to make sure your answers are correct. (Answers and corresponding problems in Arabic numerals are given for your convenience only. Do not copy these on the board.)

Example:

XIV	14	XXV	25
x III	x 3	x IV	x 4
XLII	42	C	100

LXV	65	XXVI	26
x VI	x 6	x CXI	x 111
CCCXC	390	MMDCCCLXXXVI	2,886

6. SHADOWS MEASURE HEIGHT
(Grades 4-8)

A. Preparation and Materials: You will need a stick (broomstick, etc.), yardstick and a sunny day.

B. Introduction to the Class: I wonder how high our school flagpole is? Who would like to climb it and measure it for me? (Encourage class discussion to bring out the facts that the pole is difficult to climb, it is difficult to hold a measuring device while hanging on and it is dangerous.) I will show you a way to measure the height of the flagpole without climbing it. One way would be to take this stick, place it upright on the ground and wait until its shadow is the same length as the stick. Then, we would know the shadow of the flagpole would be the same length as the flagpole. It would be easy to measure the shadow, as it is on the ground where we can get at it.

But let's suppose you wanted to know **right now.** Let's take the stick and yardstick out on the lawn and see how to do it.

You will need to find out three things. First, the length of the stick, next, the length of the stick's shadow, and last, the length of the flagpole's shadow. (Appoint one child to discover each of these facts and report to the class.)

Jean, would you please multiply the length of the stick times the length of the flagpole's shadow? (Record her answer.) Paul, would you please divide her answer by the length of the stick's shadow? (Record his answer.)

Paul's answer gives us the height of the flagpole. That was much easier than climbing the pole to measure it, wasn't it? (Encourage class discussion of other situations in which this method would be easier than measuring the actual object.)

C. Explanatory Notes: Since the angle of the sun is identical on both the stick and the flagpole, these objects, shadows and angle of sun form triangles of proportionate size. That is, the size of one triangle is in direct ratio to the size of the other.

Example:

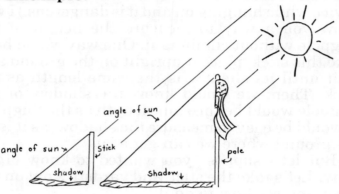

Thus the problems becomes one of simple ratio:

$$\frac{\text{Stick Shadow}}{\text{Stick Height}} : \frac{\text{Flagpole Shadow}}{\text{Flagpole Height}}$$

All numbers are known except for the flagpole height. The problem is solved in the following manner:

$$\frac{\text{Length of Stick} \quad x \quad \text{Length of Flagpole Shadow}}{\text{Length of Stick Shadow}} = \frac{\text{Height of Flagpole}}{}$$

7. TAKE A TRIP (Grades 4-8)

A. Preparation and Materials: Supply each child with a road map of the United States or of your particular state. These maps can be obtained from a service station or brought from home. Each child will also need pencil and paper.

B. Introduction to the Class: Today, each of you may take an imaginary trip. You may travel anywhere shown on your map. Study the map carefully. Decide where you would like to travel. Write down on your paper where you want to start and your destination.

Find the most direct route and trace the highways with pencil on your map. Now, on your paper write down the distance from one point to the next along the highways that you will take on your trip.

After you have the number of miles written in a column, add them to find how many miles you would have to drive to make your trip. When you have finished, we will want to hear about your trip.

C. Variation: Children could also figure out how long it would take to make the trip if they averaged fifty miles per hour and drove a maximum of four hundred miles per day. They could figure out how much gasoline they would need if their car got 20 miles to the gallon. They may wish to find out the cost of that gasoline using current local prices per gallon.

8. TIME STUDY (Grades 4-8)

A. Preparation and Materials: Have children bring a stopwatch, old magazines, time schedules for buses, trains and planes. Display in the room encyclopedias and books specifically dealing with time. Have available sheets of tagboard for mounting pictorial displays. Provide scissors, paste, staplers, etc., as needed.

You may wish to extend these suggested activities over a period of several days rather than trying to cover all of them in one class period. One period might be used to motivate action and form committee groups for independent free time study. A second period might be used for progress reports.

B. Introduction to the Class: You have been telling time since you were in early grades, but who can tell me just what time is? (The measurement of duration.) Why do you need to tell time? (To keep appointments, see specific TV programs, etc.) Do you ever need to know the number of seconds when you are telling time? (To time sports events, such as track meets, etc.) How can you tell time without a clock? (Use a sun dial, judge by the sun's position, etc.) What kinds of clocks have you seen that are not like our classroom clock? (Cuckoo, grandfather's clock, ship's clock, stopwatch, etc.)

I have a stopwatch you might enjoy using. Take turns timing how many seconds it takes to walk across the room, to read one page in a book or to work several arithmetic problems. Make a chart to record your findings.

Here are some bus, train and plane schedules. At what times do the next bus, train and plane leave for Chicago? At what time will each arrive in Chicago? Make a chart showing your findings.

There are several books on this shelf that deal with time. Who would like to make an oral report or a display chart on some of these topics related to telling time? (Allow children time to examine the books and select their own topics, such as, "What Makes A Clock Work?," "Why Do Days Grow Longer and Shorter?," "Big Ben," "Ancient

Devices for Measuring Time," "The International Date Line," etc.

Divide the class into working groups, each working in an area of personal interest. Encourage a variety of forms for final reports, such as oral reports, written reports, pictorial displays, charts and graphs or actual construction of ancient time measuring devices, etc.)

9. TIME ZONES (Grades 4-8)

A. Preparation and Materials: Draw and label the clocks as shown. Display a large map of the U.S.A. on which you have lightly marked the time zone boundaries. Have a felt pen or black crayon to darken these boundaries as the discussion indicates.

Example:

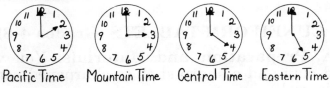

Pacific Time   Mountain Time   Central Time   Eastern Time

B. Introduction to the Class: When it is 12:00 noon in our city, why is it not noon everywhere else in the world also? Illustrate that when the sun is directly overhead in this particular spot, it is not directly over any other part of the world.

When the sun is directly over Maine, how would it appear to someone who lives in California? (By using a globe or map it would be easy to illustrate that the sun would be fairly low on the horizon over California and thus be mid-morning.)

This is why we have time zones. (The students can now draw and label time zones either on a large wall map or else on individual maps at their desks.)

I have drawn clocks showing the difference in time in our four time zones in America. How many hours difference between Pacific and Eastern time? (Compare the time differences between New York and several principal cities throughout the country.)

If a ballgame started at 2:00 in California, what time would you need to turn on your television set if you wished to watch the beginning?

(The discussion topics may lead to further research and the results could be prepared reports on such topics as Greenwich Time, Daylight Saving Time, The International Date Line, How Time Zones Affect Plane or Train Schedules, etc.)

10. CUNEIFORM TABLETS (Grades 4-8)

A. Preparation and Materials: You will need clay and a sharp stick (a sharpened pencil works well).

B. Introduction to the Class: Some of you may enjoy making cuneiform tablets like those used by the ancient Babylonians.

Flatten a piece of clay into a tablet shape. Smooth the surface carefully. Use a sharp stick or pencil to mark the cuneiform symbols. You can count one to one hundred, do mathematics problems, write your phone number or anything else you choose.

Let the tablets dry and we'll display them for everyone to enjoy.

Cuneiform Writing:

How would you add, subtract, multiply or divide using these symbols? How would you write fractions? (Lead a discussion on the difficulties involved in trying to do computational mathematics with cuneiform symbols.)

If you see " 𒐕 " does it mean one or sixty? (You can't tell unless other symbols are also given to show comparative size.) If you see " 𒌋 " does it mean ten or six hundred? (Same reasoning.)

If you had to do a lot of figuring, would you rather use our Arabic numerals or these cuneiform symbols?

11. ORIENTAL ABACUS (Grades 4-8)

A. Preparation and Materials: You will need an Oriental Abacus. Many bookstores carry them or can order one for you. You can make one with simple materials. Make a rectangular wooden frame with a crossbar 2/3 of the distance from the bottom to the top of the frame. Drill small holes through the frame and crossbar through which you can thread lengths of wire. On each wire place two small wooden beads above the crossbar, and five beads below the crossbar.

Example:

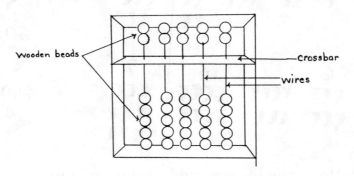

The sample exercise shown in using the abacus is long, for it includes as much variety in manipulation of the abacus as possible. Once children have mastered this exercise, they will understand the theory of the device and should be encouraged to attempt to discover for themselves how to use it to solve other types of problems and processes.

B. Introduction to the Class: Here is a device called an Oriental Abacus. It is an adding machine. This device has been in use for hundreds of years, and those skilled in its use can calculate answers faster than a person using our modern adding machines. (This has been proven in TV demonstrations.)

Each column of beads represents place value. The right hand column of beads represents the one's place, and then moving left, is the ten's place, hundred's place, and so on.

Each bead above the crossbar represents 5. Each bead below the crossbar represents 1. So to show 7 in the one's column, you could pull down 1 bead (representing 5) and push up 2 beads (each representing 1).

Example:

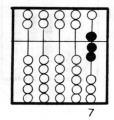

Can you show 34 on this abacus, Mary? Can you show 516, Tom?

Example:

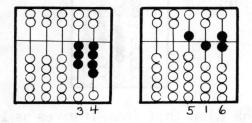

Now let's use the abacus to add 432 and 396. First, show 432 on the abacus.

Example:

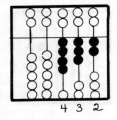

In the one's column, add 6. To do this, pull down one top bead (5) and push up 1 bottom bead.

Example:

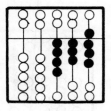

In the ten's column, add 9. There aren't 9 beads, though, are there? So watch carefully. I'll add 10 (pull down 2 top beads) and take away 1 (remove one lower bead). 10 take away 1 is the same as 9, isn't it?

Example:

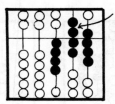

But look what that does. It gives us 12 in the ten's column. These top two beads (arrow above) stand for 10 ten's, don't they? That is the same as 1 hundred. So I will push up the 10 ten's, and move up one bead in the hundred's column. This is the same as "carrying."

Example:

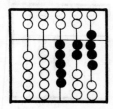

Now I will add 3 to the hundred's column. Notice the 5 single beads are already in use. But you know each top bead stands for 5, so I will push down these five bottom beads and use one top bead in their place.

Example:

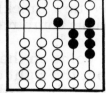

To add three, all I need to do is move up three of the lower beads in the hundred's column.

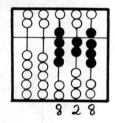

$$432$$
$$+396$$
$$\overline{828}$$

The beads show the sum of 432 and 396 to be 828. Work the problem on paper to see if this answer is correct.

You may experiment with this "adding machine" in your free time. You might see if you can figure out how to use the abacus to subtract, too.

12. CODED CARD SORTING DECK (Grades 4-8)

A. Preparation and Materials: Each child will need a 3x5 inch index card, a paper punch, scissors, ruler and pencil. You will also need a knitting needle or very long nail that will pass through the hole made by the paper punch.

B. Introduction to the Class: Have you ever seen one of those large, electronic machines that sorts cards into separate bins according to specific coded information on each card? Today, we're going to make a coded deck of cards. You'll be amazed to see how quickly we can sort them according to various combinations of coded information on each word.

First, mark a point on the top edge of the card one-half inch from the right corner. Mark a point on the right edge of the card one-half inch down from the top corner. Connect these two points. Cut along that line. This cut corner will help us keep all cards right side up and facing forward when we begin sorting.

Example:

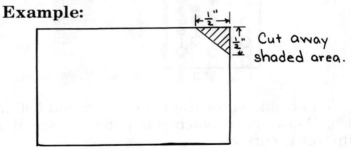

Cut away shaded area.

Now, draw a line parallel to the top of the card, and one-half inch down from the top. Beginning on the left side, mark a dot every half inch along that line. Mark a total of six dots.

Example:

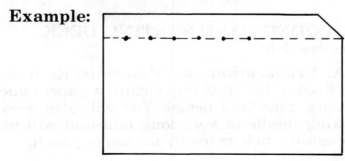

Each of you will code your card to give specific information about yourself. The cards can answer only "Yes-No" questions. So let's code the first column to answer, "Are you a boy?" If you want the card to answer "Yes," punch a hole centered over the left hand dot on your paper.

Example:

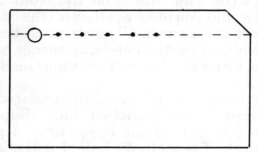

If you want the card to answer "No," first punch a hole centered over the left hand dot. Then cut away the area between the punched hole and the top of the card for "no."

Example:

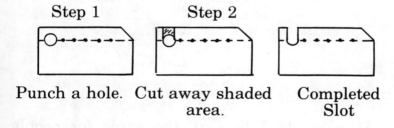

Step 1   Step 2

Punch a hole.   Cut away shaded   Completed
area.   Slot

In the next column let's code the card to answer the question, "Do you have brown hair?" If your answer is yes, punch a hole. If your answer is no, make a slot.

Continue in this way until children have coded six columns of information. The remaining questions might be:

3. Are you over 5 feet tall?
4. Do you have freckles?
5. Do you wear glasses?
6. Do you have a pet dog?

Please write your name on the front of your card. Bob, would you please collect all the cards? In order to sort these cards all cards must be facing forward. Notice how the cut-away corner helps us quickly spot any card upside down or backwards in the deck.

To sort these cards I'll use a knitting needle that passes through the punched hole. Suppose I wanted to sort out all the boys cards. I put the needle into the first-column hole. I will shake the needle briskly as I raise it up. All the cards coded "yes" have a hole in this column and they will remain on the needle. All the cards coded "no" have a slot, so they will fall off the needle.

Example:

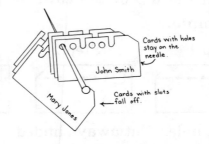

Cards with holes stay on the needle.

John Smith

Mary Jones

Cards with slots ←fall off.

How would I find all the cards for people who have a pet dog? (Put the needle in column six. Raise the needle. The "yes" cards have a hole and will stay on the needle.)

How would I find all the cards for people who do **not** have freckles? (Put the needle in column four.

Those marked "no" to freckles have a slot and will remain on the desk when the needle is raised.)

How would I find the cards for those people under five feet tall who wear glasses? (This requires two sorting steps. First, insert the needle in column three and raise it. The cards coded for under 5 feet tall have a slot and will remain on the desk. **Take those cards only** and insert the needle in column five. The cards coded "yes" for glasses have a hole and so will remain on the needle when it is lifted.)

Mary, will you please take the sorting deck and needle. How quickly can you tell me the names of all the girls with freckles? (Have her read the names when she has the cards sorted.)

John, how quickly can you find the names of all the people over 5 feet tall who do not wear glasses and who do have a pet dog? (Have him read the names when he has the cards sorted.)

Continue in this way asking such questions as:

1. Find the cards for all the boys with brown hair and freckles.

2. Find the cards for all the girls with pet dogs.

3. Find the cards for all the boys over 5 feet tall who wear glasses.

4. Find the cards for all the people without freckles who have brown hair.

See how quickly these cards can help us find any combination of facts?

13. FLOWCHARTS (Grades 4-8)

A. Preparation and Materials: The teacher will need chalk and blackboard space for illustrating the flowchart concept. Children will need pencils and paper for making their own individual flowcharts.

B. Introduction to the Class: A computer can do complicated things, but it cannot think for itself. It can only do what someone has told it to do. The person telling the computer what to do is called the **programmer.** The information he/she gives is called the **input.**

The computer can only do one step at a time, so the programmer must give the input in a series of simple steps. This step-by-step programming plan is called a **flowchart.**

Suppose each of you were a computer. I want to program you to count the books on this shelf (indicate a library shelf). My programming flowchart might look like this: (Put the following diagram on the board.)

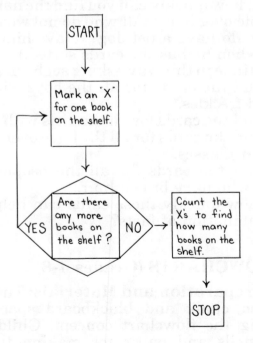

(Encourage class discussion of the breakup of procedure into a series of very simple steps.)

In this flowchart are several symbols that may be new to you. A **square box** indicates one step the computer must take. After completing that step it moves automatically to the next programmed directions.

A **diamond** indicates a two-way choice. The computer makes the choice and then moves forward on one of two alternate paths depending on which choice it has made.

An **arrow** simply indicates direction.

Notice how the computer is programmed to repeat one section of the directions while counting the books. This repeated circle of steps is called a **loop.**

Can you help me make a flowchart telling how to cross a street safely? Call on children to come to the board to diagram the step by step procedure. The completed flowchart might look like this:

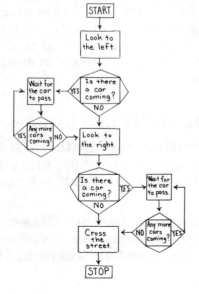

Now, let's see if you can make a flowchart of your own. Pretend you have a big bowl of ice cream and you have a spoon in your hand. Draw the flowchart that will program you to eat the ice cream.

Children's completed flowchart might look like this:

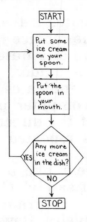

Of course, a real computer programmer must translate his flowchart into "computer language" before feeding it into the machine, but you have learned the basic concept of breaking a complicated process into a sequence of simple, step-by-step segments.

14. RANDOM SAMPLES (Grades 6-8)

A. Preparation and Materials: Cut twenty identical black squares of paper and put them in a sack. In another sack put ten black and ten white squares.

B. Introduction to the Class: Before an election, survey firms announce who probably will win. And they're almost always right. How do they know?

—218—

Survey teams take **Random Samples** of public opinion. Finding out the voting choice of **some** people helps them predict the probable voting choice of **all** people.

Here's the basic principle of random sampling. I have twenty objects in this sack. (Hold up the sack containing twenty black squares.) You cannot see what those objects are. I will let five of you come up and draw any one object each. Do not look into the sack. Pick an object at random and show the class what you have picked. (Let five children draw one object each and show the class what they drew.)

What have these people taken from my sack? (Each has an identical black square.) There were twenty things in all in this sack. You can now see just five of them. Can you guess what the rest of my objects **probably** are? (They are probably all black squares.)

Is it **possible** there is something other than black squares in this sack? (Yes.) But it is most probable that it contains all black squares. (Dump the contents onto a table so children can see the complete set.)

Now, let's take a random sample of this sack. (Hold up the sack containing ten black and ten white squares.) The complete set is twenty objects. Again you may draw out five at random. (Let five children draw one object each and show it to the class.) What do you think the rest of the objects probably are?

Poll takers do not try to survey the opinions of every single eligible voter in order to predict election results. Can you see how finding the opinions of relatively few people selected at random helps them tell the **probable** opinions of everyone?

15. COIN TOSS PROBABILITY
(Grades 6-8)

A. Preparation and Materials: Each child will need a penny, scratch paper and a pencil.

B. Introduction to the Class: We're going to work today with probability, the science of chance. Each of you has a penny. When you flip a coin, what are the two possible ways it could land, Jack? Yes, heads or tails. What determines which way it will land? (There is no way to control it. It is a matter of chance.)

On any one flip of a coin there is no way to predict exactly if that flip will turn up heads or tails. But here is how to determine what **probably** will happen in a series of flips.

Will each of you please flip your coin twenty times? Record the number of times it turned up heads and the number of times it turned up tails.

Example:

Heads	Tails
⤙⤙⤙⤙⤙ ///	⤙⤙⤙⤙⤙ ⤙⤙⤙⤙⤙ //

How many of you flipped more heads? How many flipped more tails? How many flipped an even number of each? Let's find out some of the results. (Call on five children to tell their scores, and make a chart on the board to record these scores.)

Example:

	Heads	Tails
John	6	14
Cathy	12	8
Herb	9	11
Paul	15	5
Lisa	11	9
TOTALS	53	47

Notice there is a wide variation in each person's individual scores. Some have many more heads and others have many more tails, but as we add these individual scores together notice the totals become more evenly matched.

Let's continue adding and see what happens to the totals. (Add in the results from every other child in the room. When you add the scores, you will discover the totals for heads and tails will be almost perfectly matched.)

In just a few flips of the coin no patterns emerged. But after recoding the scores for hundreds of flips you can see that heads and tails turn up with equal frequency.

You cannot predict exactly which side will turn up on any one flip of the coin, but in a series of flips it is most probable that half will be heads and half tails. On one flip chances are 50/50 it will be heads.

INDEX

CHAPTER II: "GAMES FOR ONE OR TWO"

CHAPTER III: "PUZZLES AND BRAIN TEASERS"

TABLETMASTERS™

Duplicator Masters You Tear from a Tablet!

- Faster to Use
- Easier to Handle
- Flat-Stacking

and only.......

$3⁹⁵ **each**

Factual Recall

	Grade	Cat. No.
Helps the student make use of pictures and words in a fact-finding experience, using detailed recall and supportive visual evidence. Excellent diagnostic tool for determining comprehension and ability to relate observation to written questions.	1	603-8
	2	607-0
	3	611-9

Following Directions

	Grade	Cat. No.
Provides realistic drills necessary to teach students the processes of reading, understanding, and then following directions. Each worksheet has a number of directions, each of which must be completed progressively. Verbal instructions may be added.	1	604-6
	2	608-9
	3	612-7

Thinking Skills

	Grade	Cat. No.
Study sheets call for visual sequencing, reasoning, classifying, and using cause and effect thinking skills. Sharpens each student's ability to draw conclusions, and their reasoning should be explained before answers are accepted or rejected.	1	605-4
	2	609-7
	3	613-5

Vocabulary Development

	Grade	Cat. No.
Promotes word-building through usage, exploration, similarities, expressive language and identification. Matching pictures and words reinforces the relationship between the two. A range of descriptive words, synonyms, antonyms and other forms is used.	1	606-2
	2	610-0
	3	614-3